A PRIMER *on* QUALITY *in the* ANALYTICAL LABORATORY

A PRIMER *on* QUALITY *in the* ANALYTICAL LABORATORY

John Kenkel

Southeast Community College
Lincoln, Nebraska

National Science Foundation

LEWIS PUBLISHERS

Boca Raton London New York Washington, D.C.

Library of Congress Cataloging-in-Publication Data

Kenkel, John.
A primer on quality in the analytical laboratory / John Kenkel.
    p. cm.
Includes bibliographical references and index.
ISBN 1-566-70516-9 (alk. paper)
1. Chemistry, Analytic—Quality control. 2. Chemical laboratories—Quality control. I.
Title.
QD75.4.Q34 K6   1999
543 21—dc21   99-043691

This book contains information obtained from authentic and highly regarded sources. Reprinted material is quoted with permission, and sources are indicated. A wide variety of references are listed. Reasonable efforts have been made to publish reliable data and information, but the authors and the publisher cannot assume responsibility for the validity of all materials or for the consequences of their use.

Visit the CRC Press Web site at www.crcpress.com

© 2000 by CRC Press LLC
Lewis Publishers is an imprint of CRC Press LLC

No claim to original U.S. Government works
International Standard Book Number 1-566-70516-9
Library of Congress Card Number 99-043691
Printed in the United States of America   4 5 6 7 8 9 0
Printed on acid-free paper

Preface

This work is intended to be, as the title implies, a brief introduction to the principles of quality that are important for workers in a modern industrial analytical chemistry laboratory. It is intended to be a textbook for students preparing to become technicians or chemists in the chemical process industry. It is intended to be a quick reference for new employees in an industrial laboratory as they begin to learn the intricacies of regulations and company policies relating to quality and quality assurance. It is also intended for experienced laboratory analysts who need a readable and digestible introductory guide to issues of quality, statistics, quality assurance, and regulations.

Traditionally, the education that chemists and chemistry laboratory technicians receive in colleges and universities does not prepare them adequately for some important aspects of the real world of work in their chosen field. Today's industrial laboratory analyst is deeply involved with such job issues as quality control, quality assurance, ISO 9000, standard operating procedures, calibration, standard reference materials, statistical control, control charts, proficiency testing, validation, system suitability, chain of custody, good laboratory practices, protocol, and audits. Yet, most of these terms are foreign to the college graduate and the new employee.

This book fills the void that currently exists for these individuals. It is intended to be a textbook for courses that exist or will exist in colleges and universities as teachers begin to address this gap between education and practice. But it will also be a valuable resource as new laboratory workers begin their jobs and become overwhelmed by the myriad of laboratory practices that they never learned about in school but are extremely important to their new employer.

John Kenkel
Southeast Community College
Lincoln, Nebraska

Acknowledgments

Two American Chemical Society short courses were instrumental in the development of this manuscript. These were (1) "Quality Assurance in the Analytical Testing Laboratory," taught by Gillis and Callio, and (2) "Good Laboratory Practices and ISO 9000 Standards: Quality Standards for Chemical Laboratories," taught by Mathre and Schneider.

Partial support for this work was provided by the National Science Foundation's Advanced Technological Education program through grant #DUE9751998. Partial support was also provided by the DuPont Company through their Aid to Education Program. Any opinions, findings, and conclusions or recommendations expressed in this material are those of the author and do not necessarily reflect the views of the National Science Foundation or the DuPont Company.

The author would also like to acknowledge all who read the original manuscript and/or made comments and suggestions. This list also includes those who were catalysts for the manuscript's development through their participation in a conference of chemists and chemistry technicians held at the author's institution, Southeast Community College, in 1997. The list is in alphabetical order.

John Amend, Montana State University
Clarita Bhat, Shoreline Community College
Debra Butterfield, Eastman Kodak
Steve Callio, Environmental Protection Agency
Ed Cox, Procter and Gamble
David Dellar, The Dow Chemical Company
Sue Dudek, Monsanto
Ruth Fint, DuPont
Charlie Focht, Nebraska Agriculture Laboratory
Dick Gaglione, New York City Technical College (retired)
David Hage, University of Nebraska
John Hannon, Novartis
Jim Hawthorne, DuPont
Robert Hofstader, Exxon Corporation (retired)
Kirk Hunter, Texas State Technical College
Paul Kelter, University of North Carolina — Greensboro

David Lide, Editor, CRC Handbook of Chemistry and Physics
Dennis Marshall, Eastman Chemicals
Dan Martin, LABSAF Consulting
Owen Mathre, DuPont (retired)
Ellen Mesaros, DuPont
Jane Meza, University of Nebraska
Jerry Miller, Eastman Kodak
Connie Murphy, The Dow Chemical Company
John Pederson, Dupont
Karen Potter, University of Nebraska
Reza Rafat, Pfizer
Kathleen Schulz, Sandia National Laboratory
Woody Stridde, DuPont
Richard Sunberg, Procter and Gamble
Fran Waller, Air Products and Chemicals
Gwynn Warner, Union Carbide
Carol White, Athens Area Technical Institute

Dedication

First, I dedicate this effort to my wife of 25 years, Lois, who has given me so much love for such a long time, providing such genuine happiness that it is simply overwhelming. Second, I dedicate this book to my three daughters, Angie, now known as Sister Mary Emily, and Jeanie, and Laura. In all their extraordinary goodness, I want to shout to the world what a huge blessing they are — more than any father could ever hope for. Finally, I thank my almighty Father from the bottom of my heart for giving me my faith, my family, and my talents. All good things come from Him.

Table of Contents

1 *Introduction to quality*

As citizens of the modern world and as consumers in a comfortable society, we have come to expect the highest standards of quality in all aspects of our lives. When we buy a new car, we expect that we can drive it for tens of thousands of miles free from defects in workmanship. When we elect our government officials and pay our taxes, we expect a responsive government, schools with high academic standards, air and water free of pollution, and an infrastructure that is solid and in good repair. When we pay our utility bills, we expect to always have electricity, heat, water, and a working sewer system for our homes. A quality lifestyle means excellence in consumer products, environment, health and safety, government services, and so on.

Each individual government agency and each individual private company define the terms by which the demands for quality are met within their own enterprise. A construction company will specify a particular grade of lumber in the homes it builds. A department store will stock and sell consumer products that reflect the reputation it wishes to sustain with the public relative to quality and price. A government health agency seeks to provide the health care policies and services its citizens have come to demand. A pharmaceutical company purchases raw materials, maintains a manufacturing area, hires employees, and assures the quality of its products so that it will continue to function indefinitely as a producer of drugs and medicines that the public will want to buy.

Some of these government agencies and private companies, because of the nature of their business, will utilize the services of an analytical chemistry laboratory as part of their overall need to assure the required quality operation. For example, municipal governments will employ the use of an analytical chemistry laboratory to test their water supply on a regular basis to make sure it is free of toxic chemicals. The pharmaceutical company will house an analytical chemistry laboratory within its facility to routinely test the products it produces and the raw materials that go into these products to make certain that they meet the required specifications. A fertilizer plant will utilize an analytical chemistry laboratory to confirm that the composition of its product meets the specifications indicated on the individual bags of fertilizer. Companies that produce a food product, such as snack chips, cheese, cereal, or meat products, will have an analytical chemistry laboratory as part of their operation because they want to have the assurance that the

Dr. Smith, surgeon

Figure 1.1   Given a choice, people will almost always pick quality.

products they are producing meet their own specifications for quality, consistency, and safety, as well as those of government agencies, such as the Food and Drug Administration.

In these cases, the analytical laboratory is one component of many that plays a part in a total quality scheme, or **Total Quality Management, TQM**. TQM is a concept wherein all workers within an enterprise, from upper management to custodians, are managing their own particular piece of the puzzle with utmost concern and care for quality — quality in design, quality in development, quality in production, quality in installation, and quality in servicing. Besides the laboratory, components may include manufacturing, production, research, accounting, personnel, and physical plant — virtually all aspects of an operation as depicted in Figure 1.2. The implementation of TQM emphasizes such things as (1) major paradigm shifts, if necessary, possibly meaning major cultural changes in what are routine practices and thought processes, (2) a focus on the customer, (3) a focus on improving efficiency and reducing waste, (4) a process of incorporating quality ideals in all products and processes and establishing quality criteria for all components of the enterprise, (5) a focus on training and lifelong learning, 6) a progressive management style suggesting a "team approach," (7) policies that work to identify and solve problems and constantly evaluate outcomes, 8) policies that encourage and reward employees, (9) a structure and climate conducive to quality improvement, and (10) the constructive analysis of failure. The system in place to implement TQM is often termed a **Quality System**, which consists of an organization's structure, responsibilities, procedures, and resources required for this implementation. The key lies with upper management and instillation of a positive attitude toward quality on the part of each individual employee. It then becomes a personal responsibility of each member of the team, including the laboratory personnel.

Figure 1.2   In a Total Quality Management system, all aspects of an enterprise — including managers, accountants, lab analysts, custodians, manufacturing personnel, researchers, production workers, and support staff — are focused on quality.

Laboratory personnel are as intimately involved in TQM as any other employee and aspects of their work touch on all of these ten points. The manner in which TQM principles specifically apply to laboratory personnel, however, is unique to them. They are concerned about analysis methods, choice of laboratory equipment, error analysis, statistics, acquisition of laboratory samples, etc.

How, specifically, does an analytical chemistry laboratory assure the quality of its work? The purpose of this monograph is to discuss the processes utilized by analytical chemistry laboratories through which the results reported to their customers and clients, whether internal to their company or external, are assured to be of the highest quality and greatest accuracy possible. The methods, procedures, and techniques employed by these laboratories for the individual analyses that they perform are what are called into question and tested. In most cases, methods of statistics must be applied because the measurement techniques are subject to errors that often cannot be identified or compensated.

2 Quality standards and regulations

In today's world, the economies of nations are intertwined. Raw materials mined or manufactured in one country are sold in another. Industrial, agricultural, and other products manufactured in one country are sold in another. In the U.S., foreign products from automobiles to toys are commonplace. The American farmer sells grain to other countries. Middle Eastern countries sell crude oil to the U.S. and other countries. The list is long, and thus the demand for quality is global.

For this reason, an international standards organization governing global quality has been created. It is called the **International Organization for Standardization or ISO**, and is a worldwide federation of national standards bodies. The purpose of the organization is to promote common standards developed by its technical committiees. Each member body has a right to be represented on a committee. The U.S. member body is called the **American National Standards Institute, or ANSI**. In turn, the **American Society for Quality, or ASQ**, is the U.S. member of ANSI responsible for quality management and related standards. The ISO standards are generic and apply to any industry (Figure 2.1).

The current set of quality standards endorsed by the ISO is the **ISO 9000** series. This series is a set of documents drafted by the member delegates and is intended primarily to ensure that the exchange of goods between companies is of high and internationally acceptable quality. ANSI and ASQ have adopted ISO 9000 word-for-word for use in the U.S. The original documents, published in 1994, are designated ISO 9000, ISO 9001, ISO 9002, ISO 9003, and ISO 9004. The corresponding ANSI/ASQ designations are ANSI/ASQ Q9001-1994 through ANSI/ASQ Q9004-1994. While the ISO standards address quality management and quality assurance, they do not provide test methods or quality control procedures for laboratories. However, ISO, in conjunction with the International Electrotechnical Commission (IEC), has published ISO/IEC Guide 25, which lists the general requirements for the competence of calibration and testing laboratories. The ISO series is important because it can be the basis by which laboratories, indeed entire

ORGANISATION
INTERNATIONALE DE
NORMALISATION

INTERNATIONAL
ORGANIZATION FOR
STANDARDIZATION

Figure 2.1   Official logos for ISO and ANSI, the two organizations that impact quality in the U.S.

companies, become internationally registered, accredited, and/or certified. ISO 9000 certification will be discussed in Section 9.

The ISO has also produced a set of quality standards specifically for environmental management. This is the ISO 14000 series. The areas addressed by ISO 14000 are Environmental Management Systems, Environmental Performance Evaluations, Environmental Auditing, Life Cycle Assessment, and Environmental Labeling.

Besides ISO standards, pharmaceutical companies in the U.S. are governed under certain circumstances by separate federal regulations adopted by the Food and Drug Administration (FDA). These regulations are known as **Current Good Manufacturing Practices**, or **cGMP**. The cGMP were developed to ensure that pharmaceutical products are produced and controlled according to the quality standards pertinent to their intended use. The cGMP are found in Parts 210 and 211 of Chapter 21 of the Code of Federal Regulations (21 CFR 210 and 21 CFR 211). Also, U.S. environmental laboratories, pharmaceutical laboratories, and laboratories found within chemical companies in the U.S. are governed by separate federal regulations adopted by the Environmental Protection Agency (EPA) as well as the FDA. These regulations are known as **Good Laboratory Practices**, or **GLP**. The EPA GLPs are found in Part 160 of Chapter 40 of the CFR (40 CFR 160) and the FDA GLPs are found in Part 58 of Chapter 21 of the CFR (21 CFR 58). GLP will also be addressed in detail in Section 7.

3 Principles and terminology of quality assurance

First, one should distinguish between quality assurance and quality control. **Quality control** can be defined as the overall system of operations designed to control a process so that a product or service adequately meets the needs of the consumer. **Quality assurance** is the system of operations that tests the product or service to ensure compliance with defined specifications. In a candy factory, quality control would consist of the company procedures to ensure that the candy-making process is set up to be free of potential contamination sources, such as insects, hair strands, etc., while the quality assurance operations might simply consist of a random tasting of the product. For a company that manufactures basketball hoop and backboard units, the quality control operation might consist of regular inspection of the manufacturing operation and its components and processes, such as the welding process, to see that it is being carried out according to specification. Quality assurance would consist of a random testing of the finished products for strength, proper dimensions, etc. In an analytical chemistry laboratory, a quality control program would consist of the system in place to monitor the overall performance. Are the lab workers properly trained? Are the instruments properly calibrated? Are the key elements of the program being properly documented? A quality assurance operation consists of the laboratory testing of the company products, or an agency's samples, etc., to determine if they are within specification.

Consider what is termed the **sample**. A sample is a small portion of a large bulk system that is acquired and taken into the analytical laboratory in lieu of the entire system. For example, it is not practical to bring the entire contents of a 5000-gallon tank of liquid sugar solution used in a pharmaceutical preparation into the analytical laboratory for analysis. The small portion of the solution, perhaps a small vial, that is obtained for the analysis is called the sample, and that is what is taken into the laboratory for analysis. How well a sample represents the entire bulk system, and what fundamental issues of quality are involved when obtaining the sample are important questions and will be dealt with in Section 6. The process of obtaining the sample is referred to as **sampling**. The component of the sample that is under investigation, and for which a concentration level is sought, is call the **analyte**.

The measurements made and results reported on the sample must be **valid**. This means that the sampling and measurement systems must be perfectly applicable to the system under investigation, the instruments and measuring devices used must be **calibrated**, and the data must be handled and the results must be calculated and/or reported according to nominally acceptable norms that are well grounded in scientific principles and facts. Accordingly, all sampling, measurement, and reporting schemes proposed or used for a given analysis must be **validated**, and it is often the full-time job of one or more experienced laboratory analysts to perform this **validation study**. To a certain extent, this work is a research project. After a new method is proposed for a given work, the analyst must execute the procedure repeat-edly using a sample with an expected outcome in order to gather information relating to precision, accuracy, and bias. These latter terms are defined below.

The **measurement system** mentioned above consists of all the physical equipment, facilities, logistics, and processes that need to be configured in order to make the measurement that is needed. These can include sampling locations (from what parts of the whole bulk system does one take a sample), the actual taking of the sample (equipment and technique), the laboratory preparation of the sample, the instruments and equipment needed in the laboratory, and **calibration** and data handling methods. **Accuracy** is the degree to which the result obtained agrees with the correct answer. (Usually, the correct answer is not known.) **Precision** is the degree to which a series of measurements made on the same sample with the same measurement system agree with each other. **Bias** is an error that occurs over and over again (systematic) due to some fault of the measurement system. Precision, accuracy, and bias are illustrated in Figure 3.1.

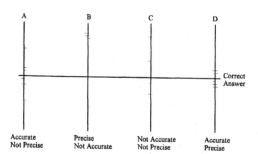

Figure 3.1   An illustration of precision, accuracy, and bias. When accurate but not precise, the measurements are bunched loosely around the correct answer (A). When measurements are bunched, but not around the correct answer, they are precise but not accurate, and a bias is indicated (B). When there is a large spread in the mea-surements and the mean is not near the correct answer, they are neither precise nor accurate (C). When accurate and precise, the measurements are bunched tightly around the correct answer (D).

Calibration is a procedure by which an instrument or measuring device is tested in order to determine what its response is for an analyte in a test sample for which the true response is either already known or needs to be established.   One then either makes an adjustment so that the known response is, in fact, produced, correlates the response of unknowns with that of the known quantity, or, if the device or instrument is deemed defective, either removes the device from service permanently or effects repairs. For example, when calibrating a pH meter, one immerses the pH probe into a test solution whose pH is known (buffer solution) and then tweaks the electronics so that it gives that pH on the display. When calibrating a balance, one places an object of known weight on the pan. If the correct weight is displayed, the balance is calibrated for that weight of sample. If the correct weight is not displayed, one concludes that the balance is out-of-calibration and it is taken out of service. When calibrating a spectrophotometer, one measures the instrument's response for a series of known test samples, all of a different concentration, and plots the response vs. concentration (a so-called **calibration curve** or **standard curve**; see Figure 3.2.). If it is linear, the instrument is said to be calibrated and unknown samples can be correlated with their responses to give the results.

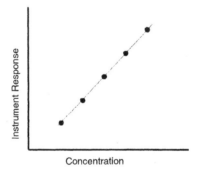

Figure 3.2   A calibration curve or standard curve.

At the end of Section 1, it was mentioned that measurement techniques are subject to errors, and bias was also mentioned. In general, errors are of three types: (1) those that are **systematic errors** and produce a known bias in the data, (2) those that are avoidable blunders that are known to have occurred, or were found later to have occurred, the so-called **determinate errors**, and (3) those called **random errors**, or also **indeterminate errors**, which are errors that occur, but can neither be identified nor directly compensated. Correction factors can be applied to data resulting from systematic errors. Measurements resulting from determinate errors can be discarded. Random errors are dealt with by applying concepts of statistics to the data. Section 4 will deal with this very important aspect of quality assurance in an analytical laboratory.

4 Elementary statistics

4.1 Introduction

Accuracy in the laboratory is obviously an important issue. If the analysis results reported by a laboratory are not accurate, everything a company or government agency strives for, the entire TQM system, may be in jeopardy. If the customer discovers the error, especially through painful means, the trust the public has placed in the entire enterprise is lost. For example, if a baby dies due to nitrate contamination in drinking water that a city's health department had determined to be safe, that department, indeed the entire city government, is liable. In this "worst-case scenario," some employees would likely lose their jobs and perhaps even be brought to justice in a court of law.

As noted in the last section, the correct answer to an analysis is usually not known in advance. So the key question becomes: **How can a laboratory be absolutely sure that the result it is reporting is accurate?** First, the bias, if any, of a method must be determined and the method must be validated as mentioned in the last section (see also Section 5.6). Besides periodically checking to be sure that all instruments and measuring devices are calibrated and functioning properly, and besides assuring that the sample on which the work was performed truly represents the entire bulk system (in other words, besides making certain the work performed is free of avoidable error), the analyst relies on the precision of a series of measurements or analysis results to be the indicator of accuracy. If a series of tests all provide the same or nearly the same result, and that result is free of bias or compensated for bias, it is taken to be an accurate answer. Obviously, what degree of precision is required and how to deal with the data in order to have the confidence that is needed or wanted are important questions. The answer lies in the use of **statistics**. Statistical methods take a look at the series of measurements that are the data, provide some mathematical indication of the precision, and reject or retain **outliers**, or suspect data values, based on predetermined limits.

4.2 Definitions

Some definitions that are fundamental to statistical analysis include the following.

Mean: In the case in which a given measurement on a sample is repeated a number of times, the average of all measurements is an important number and is called the *mean*. It is calculated by adding together the numerical values of all measurements and dividing this sum by the number of measurements.

Median: For this same series of identical measurements on a sample, the "middle" value is sometimes important and is called the *median*. If the total number of measurements is an even number, there is no single "middle" value. In this case, the median is the average of two "middle" values. For a large number of measurements, the mean and the median should be the same number.

Mode: The value that occurs most frequently in the series is called the *mode*. Ideally, for a large number of identical measurements, the mean, median, and mode should be the same. However, this rarely occurs in practice. If there is no value that occurs more than once, or if there are two values that equally occur most frequently, then there is no mode.

Deviation: How much each measurement differs from the mean is an important number and is called the *deviation*. A deviation is associated with each measurement, and if a given deviation is large compared to others in a series of identical measurements, this may signal a potentially rejectable measurement (outlier) which will be tested by the statistical methods. Mathematically, the deviation is calculated as follows:

$$d = |m - e| \qquad (4.1)$$

in which d is the deviation, m is the mean, and e represents the individual experimental measurement. (The bars (| |) refer to "absolute value," which means the value of d is calculated without regard to sign; i.e., it is always a positive value.)

Sample Standard Deviation: The most common measure of the dispersion of data around the mean for a limited number of samples (<20) is the sample standard deviation:

$$s = \sqrt{\frac{d_1^2 + d_2^2 + d_3^2 + \dots}{n - 1}} \qquad (4.2)$$

The term $(n - 1)$ is referred to as the number of **degrees of freedom**, and s represents the standard deviation.

Example 4.1

The percent moisture in a powdered pharmaceutical sample is determined by six repetitions of the Karl Fisher method to be 3.048%, 3.035%, 3.053%, 3.044%, 3.049%, and 3.046%. What are the mean, median, mode, and sample standard deviation for these data?

Solution 4.1

The mean is the average of all measurements. Thus, one has:

$$\text{Mean} = \frac{(3.048 + 3.035 + 3.053 + 3.044 + 3.049 + 3.046)}{6}$$

$$= 3.045833 = 3.046\%$$

The median is the "middle" value of an odd number of values. If there is an even number, the median is the average of the two "middle" values. Thus, one has:

$$\frac{3.048 + 3.046}{2} = 3.047\%$$

There is no mode in this case because there is no value appearing more than once.

The sample standard deviation is calculated according to Equation 4.2, in which the "d" values are deviations calculated by subtracting each individual percent value from the mean according to Equation 4.1. The deviations (absolute values) are 0.002, 0.011, 0.007, 0.002, 0.003, and 0.000. To substitute into Equation 4.2, one must square the deviations. The squares of the deviations are 0.000004, 0.000121, 0.000049, 0.000004, 0.000009, 0.000000. Substituting into Equation 4.2, one obtains:

$$s = \sqrt{\frac{d_1^2 + d_2^2 + d_3^2 + \ldots}{n - 1}}$$

$$= \sqrt{\frac{0.000004 + 0.000121 + 0.000049 + 0.000004 + 0.000009 + 0.000000}{(6 - 1)}}$$

$$= 0.0061155 = 0.006$$

The significance of the sample standard deviation is that the smaller it is numerically, the more precise the data and thus presumably (if free from bias and determinate error) the more accurate the data.

Population Standard Deviation: The dispersion of data around the mean for the entire population of possible samples (an infinite number of samples), which is approximated by $n > 20$, is called the *population standard deviation* and is given the symbol σ (Greek letter sigma).

$$\sigma = \sqrt{\frac{d_1^2 + d_2^2 + d_3^2 + \dots}{n}} \tag{4.3}$$

Variance: A more statistically meaningful quantity for expressing data quality is the variance. For a finite number of samples, it is defined as:

$$v = s^2 \tag{4.4}$$

It is considered to be more statistically meaningful because if the variation in the measurements is due to two or more causes, the overall variance is the sum of the individual variances. The variance for Example 4.1 is $(0.0061155)^2$, or 0.0000037, or 0.000004. For an entire population of samples, s is replaced by σ.

Relative Standard Deviation: One final deviation parameter is the relative standard deviation, RSD. It is obtained by dividing s by the mean.

$$\text{RSD} = \frac{s}{m} \tag{4.5}$$

Multiplying the RSD by 100 gives the *relative % standard deviation*:

$$\text{Relative \% standard deviation} = \text{RSD} \times 100 \tag{4.6}$$

Multiplying the RSD by 1000 gives the relative parts per thousand (ppt) standard deviation:

$$\text{Relative ppt standard deviation} = \text{RSD} \times 1000 \tag{4.7}$$

The relative % standard deviation (Equation 4.5) is also called the **coefficient of variance**, c.v. Relative standard deviation relates the standard deviation to the value of the mean and represents a practical and popular expression of data quality. Again, for an entire population of samples, s is replaced by σ.

Example 4.2

What is the %RSD for the data in Example 4.1?

Solution 4.2

The %RSD is calculated according to Equation 4.6, or:

$$\%\text{RSD} = \frac{s}{m} \times 100 = \frac{0.0061125}{3.045833} \times 100 = 0.2\%$$

4.3 Distribution of Measurements

If the entire universe of data (the **population**), as opposed to just a small number of samples, is graphically displayed in a plot of frqeuency of occur-rence vs. individual measurement values, a bell-shaped curve would result in which the peak of the curve would coincide with the mean, as shown in Figure 4.1. This graph is called the **normal distribution curve**. It shows that, for an entire population, the measurements are dispersed around the mean with an equal "drop-off" from the mean in each direction. This mean is recognized as the **true mean** because the entire population was analyzed. The true mean is designated with "μ" (Greek letter mu). The **population standard deviation** is associated with μ and, as indicated in Section 4.2, is designated as σ. A bias can be depicted on a normal distribution curve by drawing a verticle line at the position of the correct answer to an analysis (see Figure 4.2). In addition, the concept of precision can also be depicted. The more precise the data, the tighter the data points are bunched around the mean and the smaller the σ. The less precise, the less tight the data and the larger the σ (see Figure 4.3).

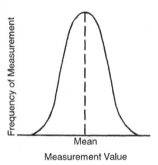

Figure 4.1 The normal distribution curve.

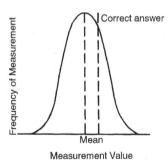

Figure 4.2 A normal distribution curve displaced from the correct answer due to a bias.

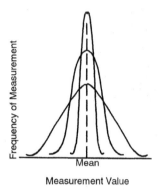

Figure 4.3   Several normal distribution curves superimposed to illustrate variations in precision.

On a normal distribution curve, 68.3% of the data falls within one σ on either side of the mean, 95.5% of the data falls within two σ on either side of the mean, and 99.7% of the data falls within three σ on either side of the mean. Figure 4.4 shows the 2σ limits on either side of the mean.

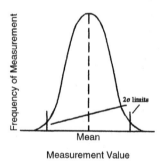

Figure 4.4   A normal distribution curve with the 2σ limits indicated.

For a small number of samples, it is sometimes useful to plot a histogram of the data in order to pictorially show the distribution. A **histogram** is a bar graph that plots ranges of values on the *x*-axis and frequency on the *y*-axis. An example is shown in Figure 4.5. Each vertical bar represents a range of measurement values on the *x*-axis. The height of each bar represents the number of values falling within each range.

4.4   *Student's* t

In order to express a certain degree of confidence that the mean deter-mined in a real data set is the true mean, **confidence limits** are established

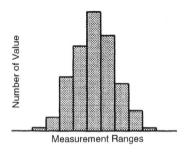

Figure 4.5 An example of a histogram for a finite number of measurements.

based on the degree of confidence, or **confidence level**, that the analyst wishes to have for the analysis. The confidence limit is the interval around the mean that probably contains the true mean, μ. The confidence level is the probability (in percent) that the mean occurs in a given interval. A 95% confidence level means that the analyst is confident that for 95% of the tests run, the sample will fall within the set limits.

The more measurements made on a given bulk system, the more the histogram in Figure 4.5 would begin to look like the normal distribution curve in Figure 4.1. In other words, the more measurements made, the closer the value of the mean will be to the true mean, and the more we can rely on the mean to be the correct answer in the absence of bias. Similarly, the more measurements made, the closer the value of the standard deviation is to the population standard deviation. From a practical point of view, however, one only runs an experiment enough times to provide the confidence interval desired. The confidence interval represents the range from the lower confidence limit to the upper confidence limit. For example, for a mean of 23.54, if the confidence limits are 23.27 and 23.81 (0.27 on either side of the mean) the confidence interval would be ± 0.27. To express the degree of confidence in the mean, the answer to the analysis, or what could be called the "true mean," could then be expressed as 23.54 ± 0.27.

A statistically appropriate way of determining the confidence interval for a desired confidence level is the Student's *t* method. This method expresses the true mean as follows:

$$\mu = m \pm \frac{ts}{\sqrt{n}} \tag{4.8}$$

in which *t* is a constant depending on the confidence level, and *n* is the number of measurements. The values of *t* required for a desired confidence level and for a given number of measurements are given in Table 4.1. (See also Box 4.1.)

Table 4.1 Student's *t* Values for Various
Confidence Levels and Numbers of Measurements

	Confidence Level		
n − 1	90	95	99
1	6.314	12.706	63.657
2	2.920	4.303	9.925
3	2.353	3.182	5.841
4	2.132	2.776	4.604
5	2.015	2.571	4.032
6	1.943	2.447	3.707
7	1.895	2.365	3.500
8	1.860	2.306	3.355
9	1.833	2.262	3.250
10	1.812	2.228	3.169
∞	1.645	1.960	2.576

Example 4.3

What is the true mean (with confidence interval) for
the data in Example 4.1 using Student's *t* for the 95%
confidence level?

Solution 4.3

Using Equation 4.6 and the value of *t* from Table 4.1
for a confidence level of 95%, one obtains:

$$\mu = 3.046 \pm \frac{2.571 \cdot 0.0061125}{\sqrt{6}} = 3.046 \pm 0.006$$

4.5 Rejection of Data

It may at times appear that a single measurement (an outlier) is so different
from the others that the analyst wonders if there was some determinate error
that was not detected. In that case, a decision must be made as to whether
this measurement should be "rejected," meaning not included in the calculation of the mean. This measurement should not be immediately rejected
as being "bad" because, in the absence of a full investigation to determine
a cause, it may, in fact, be legitimate. If a legitimate measurement is rejected,
then a bias is introduced, and the mean, while assumed to be the correct
answer, actually is flawed. There must be some criterion adopted for the
rejection or retention of such data.

The first course of action would be for the analyst to inspect his/her
technique, chemicals, notebook records, and perhaps the equipment used,
to try to detect a determinate error. If a cause is found, then the measurement
should be rejected and the reasons for such rejection documented. If a cause
is not found, and if time is not a factor, it would be advisable to repeat the

The Story of Student

"For a long time, many investigators did not attempt to make any statement about the average, particularly if the average was based on very few measurements. The mathematical solution to this problem was first discovered by an Irish chemist who wrote under the pen name "Student." Student worked for a company that was unwilling to reveal its connection with him lest its competitors would discover that Student's work would also be advantageous to them. It now seems extraordinary that the author of this classic paper on measurements was not known for more than 20 years. Eventually it was learned that his real name was William Sealy Gosset (1876–1937)."

From Youdon, W.J., *Experimentation and Measurement*, National Institute of Standards and Technology Special Publication 672, 1961, reprinted 1997.

Box 4.1   The story of Student, as related by W.J. Youdon in the publication cited.

measurement, perhaps many times, to see if the anomaly appears again. If it does, the situation is not resolved unless a cause is established in the course of the repetition. If it does not appear again, its seriousness has diminished because there are more measurements from which the mean is calculated.

For small data sets ($n < 10$), which are often encountered in chemical analysis, a simple method to determine if an outlier is rejectable is the **Q test**. In this test, a value for Q is calculated and compared to a table of Q values that represent a certain percentage of confidence that the proposed rejection is valid. If the calculated Q value is greater than the value from the table, then the suspect value is rejected and the mean recalculated. If the Q value is less than or equal to the value from the table, then the calculated mean is reported. Q is defined as follows:

$$Q = \frac{\text{Gap}}{\text{Range}} \tag{4.9}$$

where the "gap" is the difference between the suspect value and its nearest neighbor, and the "range" is the difference between the lowest and highest values. A list of Q values for the 90% confidence level is given in Table 4.2.

Table 4.2 Values of Q at the 90% Confidence Level for Different Numbers of Measurements

Number of Measurements	Q (at 90% confidence level)
3	0.94
4	0.76
5	0.64
6	0.56
7	0.51
8	0.47
9	0.44
10	0.41

Example 4.4

Determine if any of the values in Example 4.1 should be rejected based on the Q test at the 90% confidence level.

Solution 4.4

The six values re-aligned from lowest to highest are 3.035%, 3.044%, 3/046%, 3.048%, 3.049%, and 3.053%. The outlier would be 3.035%, since it has a larger gap (0.009) from its nearest neighbor (3.044) than has 3.053. Thus, one obtains:

$$Q = \frac{Gap}{Range} = \frac{0.009}{0.018} = 0.5$$

Since the Q value for 6 measurements is 0.56 (Table 4.2), and since the calculated value (0.5) is less, the suspect value cannot be rejected.

4.6 Final Comments on Statistics

When chemists talk about an analytical method or when instrument vendors tout their products, they often quote the standard deviation that is achievable with the method or instrument as a measure of quality. For example, the manufacturer of an HPLC pump may declare that the digital flow control for the pump, with flow rates from 0.01 to 9.99 mL per minute, has a RSD less than 0.5%, or a chemist declares that her atomic absorption instrument gives results within 0.5% RSD. The most fundamental point about standard deviation is that the smaller it is, the better, because the smaller it is, the more precise the data (the more tightly bunched the data are around the mean) and, if free of bias, the greater the chance that the data are more accurate. Chemists have come to know through experience that a 0.5% RSD for the flow controller and, under the best of circumstances, a 0.5% RSD for atomic absorption results are favorable RSD values compared to other comparable instruments or methods.

5 The practice of quality assurance

5.1 Introduction

In Section 3, quality assurance was defined as the laboratory operations employed to test a company's products, or an agency's samples, etc., to determine if they are within specification. This section discusses the specifics of these operations — what considerations are involved in the day-to-day work of a quality assurance technician or chemist.

5.2 Standard Operating Procedures

The documented set of instructions a technician or chemist follows when carrying out an analysis or process is called the **Standard Operating Procedure**, or **SOP**. SOPs are very carefully considered instructions that become official only after thorough review and testing by the laboratory personnel. Careful attention to such written instructions is required in order for the work to be considered valid under GLP regulations. The concept of an SOP, and what areas require the use of SOPs, are addressed in nearly identical statements in the EPA GLP regulations, 40 CFR 160.81, and in the FDA GLP regulations, 21 CFR 58.81. The EPA text is reproduced here in Box 5.1. An SOP is intended to ensure the quality and integrity of the data generated in a laboratory.

SOPs can be both general and specific. Examples of general laboratory operations include how to characterize an analytical standard, how to record observations and data, and how to label reagents and solutions. Most laboratory operations even have an SOP for writing and updating SOPs. Examples of specific laboratory operations include the preparation and analysis of a specific company's product or raw material, the operation and calibration of specific instruments, and the preparation of specific samples for analysis. Often, SOPs are based on published methods, such as those found in scientific journals, in application notes, and procedures published by instrument manufacturers, or in books of standard methods, such as those published by the **American Society for Testing and Materials (ASTM)** and the **Association of Official Analytical Chemists (AOAC)**. The published

---

Standard Operation Procedures as quoted from 40 CFR 160.81

(a) A testing facility shall have standard operating procedures in writing setting forth study methods that management is satisfied are adequate to ensure the quality and integrity of the data generated in the course of a study. All deviations in a study from standard operating procedures shall be authorized by the study director and shall be documented in the raw data. Significant changes in established standard operating procedures shall be properly authorized in writing by management.

(b) Standard operating procedures shall be established for, but not limited to, the following:
   (1) Test system area preparation.
   (2) Test system care.
   (3) Receipt, identification, storage, handling, mixing, and method of sampling of the test, control, and reference substances.
   (4) Test system observations.
   (5) Laboratory or other tests.
   (6) Handling of test systems found moribund or dead during study.
   (7) Necropsy of test systems or post-mortem examination of test systems.
   (8) Collection and identification of specimens.
   (9) Histopathology.
   (10) Data handling, storage and retrieval.
   (11) Maintenance and calibration of equipment.
   (12) Transfer, proper placement, and identification of test systems.

(c) Each laboratory or other study area shall have immediately available manuals and standard operating procedures relative to the laboratory or field procedures being performed. Published literature may be used as a supplement to standard operating procedures.

(d) A historical file of standard operating procedures, and all revisions thereof, including the dates of such revisions, shall be maintained.

Box 5.1   The concept of the SOP, as addressed in the EPA GLP, 40 CFR 160.81.

methods are only resources, however, since they must be adapted to specific samples, equipment, etc. and must follow a specific format under GLP regulations. It is very important that SOPs be adequate to ensure the quality and integrity of the data to be obtained.

When a new SOP is to be written or when an outdated one is to be revised, some very important protocols must be considered. First, the scientists who will be conducting the work must be involved. They are most familiar with the laboratory facilities and equipment and can most likely provide very useful input. Second, a standard format must be followed so that all SOPs look alike and provide information the company employees and laboratory auditors expect. The format typically includes a document number, a descriptive title, a revision number, an effective date, a statement of purpose, a statement of scope, the procedure itself, references (such as the ASTM or AOAC references), and also the required company signatures. An example is given in Box 5.2.

Policies and Procedures for Laboratory Notebooks and Data Recording SOP NB-D
Revision #4   Replaces NB-C   Effective Date: November 1, 1998
Scope: This operating procedure applies to the use of all laboratory notebooks and log books used in the laboratories.
Purpose: The purpose of this operating procedure is to ensure uniformity among all laboratories so that the quality and traceability of all data recording are assured.
Prepared by: <u>John Kenkel</u>   Approved by: <u>David Newton</u>

I. General Guidelines
A. All notebooks must begin with a Table of Contents. All pages must be numbered and these numbers must be referenced in the Table of Contents. The Table of Contents must be updated as projects are completed and new projects begun.
B. All notebook entries will be made in black ink. Use of graphite pencils or other erasable writing instrument is strictly prohibited.
C. No entries will be erased or otherwise made illegible. If an error was made, a single line will be drawn through the entry. Do not use correction fluid. Initial and date corrections and indicate why the correction was necessary.
D. Under no circumstances will any notebook be taken home or otherwise leave the laboratory unless there are data to be recorded at a remote site, such as at a remote sampling site or unless special permission is granted by the supervisor.
E. The following notebook format should be maintained for each project undertaken: 1) Title and Date; 2) Purpose or Objectives Statement; 3) Data Entries; 4) Results; 5) Conclusions. Each of these are explained below.
F. Make notebook entries for a given project on consecutive pages where practical. Begin a new page for each new project. Do not skip pages.
G. Draw lines through blank spaces or pages. These spaces or pages should be initialed and dated.
H. Never use a highlighter in a notebook.
I. Each notebook page must be signed and dated by the person doing the work.
II. Title and Date
A. All new experiments will begin with the title of the work and the date it is performed. If the work was continued on another date, that date must be indicated at the point the work was re-started.
B. The title will reflect the nature of the work.
III. Purpose or Objectives Statement
A. Following the Title and Date, a statement of the purpose or objective of the work will be written. This statement should be brief and to the point.
B. If appropriate, the Standard Operating Procedure will be referenced in this statement.
IV. Data Entries
A. Enter data into the notebook as the work is being performed. Entries should be made in black ink only.
B. If there is any deviation from the SOP, this must be thoroughly documented by indicating exactly what the deviation was and why it occurred.
C. The samples analyzed must be described in detail. Such descriptions may include the source of the sample, what steps were taken to ensure that it represents the whole (reference SOP if appropriate) and what special coding may be assigned and what the codes mean. If the codes were recorded in a separate notebook (such as a "field" notebook), this notebook must be cross-referenced.
D. Show the mathematical formula utilized for all calculations and also a sample calculation. Computer programs used for data analysis should be referenced.
E. Construct data "tables" whenever useful and appropriate.
F. Both numerical data and important observations should be recorded.
G. Limit attachments (chart recordings, computer printouts, etc.) to one per page. Clear tape or glue may be used. Do not use staples. Only one fold in attachments is allowed. Do not cover any notebook entries with attachments.
V. Results
A. The results of the project, such as numerical values representing analysis results, should be reported in the notebook in table form if appropriate. Otherwise, a statement of the outcome is written, or if a single numerical value is the outcome, then it is reported here.
VI. Conclusion
A. After results are reported, the experiment is drawn to a close with a brief concluding statement indicating whether the objective was achieved.

Box 5.2   An example SOP.

All current SOPs should be available in the work area in which they are used. Each person who may need specific SOPs for his/her work should also have them, perhaps in a file near his/her desk. In addition, there should be a location in which master SOPs for all activities are filed and all SOPs should also be archived so that past revisions are accessible. All obsolete SOPs, however, should be removed and filed away from the work area and clearly identified as obsolete. The decision to revise an SOP must be based on sound observations and protocols that point to improved data accuracy and integrity. Such decisions can be based on a new procedure, a new piece of equipment, etc. SOPs are dynamic documents and should be considered for revision on a regular basis with input from the technicians and scientists doing the work.

Deviating from a current SOP is important to consider. The technician or scientist doing the work should not change a procedure at will and certainly not without proper documentation and discussion. Deviations can be authorized by management and by a study director, but even then, not without proper explanation and documentation.

5.3   Calibration and Standardization

Both the ISO 9000 guidelines (ISO/IEC Guide 25 — see Section 2) and the GLP regulations (e.g., FDA regulations 21 CFR 58) stress the importance of the calibration of laboratory equipment. (See Box 5.3.) Indeed, properly calibrated test equipment is central to a laboratory's duty to successfully and accurately carry out its responsibilities.

In Section 3, **calibration** was defined as a procedure by which an instrument or measuring device is tested in order to determine what its response is for an analyte in a test sample or samples for which the true response is either already known or needs to be established. If the true response is already known, one then makes an adjustment, if possible, so that the known response is, in fact, produced. If one cannot adjust to give the known response, the device is defective and is taken out of service and repaired. If the true response needs to be established, one establishes it via a single

---

ISO/IEC Guide 25 Statement

"All measuring and test equipment having an effect on the accuracy or validity of calibrations or tests shall be calibrated and/or verified before being put into service. The laboratory shall have an established program for calibration and verification of its measuring and test equipment."

FDA Regulation 21 CFR 58 (GLP) Statement

"Equipment used for the generation, measurement, or assessment of data shall be adequately tested, calibrated, and/or standardized."

Box 5.3   Statements from the ISO and GLP documents that guide and regulate analytical laboratories.

standard, or perhaps via a calibration curve or standard curve created using a series of standards, and then correlates the response of unknowns with that of the known quantity or quantities.

Calibration is a very critical component of quality control practices. If the measuring devices are not giving the response for a standard that they are expected to give, they cannot be expected to give an accurate response for an unknown sample. Uncalibrated equipment ensures an inaccurate result.

A sample for which the true response is already known or is established is called a **standard**. A standard can be a **primary standard**, which is a standard through which other substances or solutions are made to be standards. It can also be a **secondary standard**, a solution whose concentration is known accurately either because it was prepared using a primary standard or because it was compared to another standard. All standards must ultimately be traced to a **standard reference material (SRM)**. Standard reference materials are available from the **National Institute of Standards and Technology (NIST)** and should not be used for any other purpose in the laboratory (Section 5.4). **Standardization** is an experiment in which a solution is compared to a standard in order for itself to be a standard. The solutions used to establish a standard curve are often called **reference standards** and these must also be traceable to an SRM.

Some examples of calibration and standardization are presented below.

Volumetric Glassware: The graduation lines on volumetric glassware, such as glass flasks (Figure 5.1), pipets (Figure 5.2), and burets, are permanently affixed on each piece at the factory. In other words, they are calibrated at the factory, and this calibration cannot be adjusted nor can it be changed unless the item is mistreated in some way, such as scratched by the brushes used for cleaning, etched by the use of chemical agents, or heated in an oven. Of course, cleanliness is also an issue, and tips of pipets and burets must not be chipped so as to change the delivery time of the fluid. If the item is designated as Class A, the manufacturer guarantees that such calibration meets the specifications

Figure 5.1   Volumetric flask.

Figure 5.2   Pipet.

of NIST and the calibration does not need to be checked in the laboratory unless the highest degree of accuracy is required. Glassware that is not Class A should be used only when accuracy is less important.

It is possible to check the calibration of a pipet, flask, or buret. The process involves weighing with a calibrated analytical balance. The volume of water (temperature noted) delivered or contained by the glassware is weighed. Then the analyst converts this weight to volume (using the density of water at the temperature noted), corrects the result to 20°C (the usual temperature of the factory calibration), and compares it to the factory calibration. If the difference is not tolerable, the piece of glassware is either not used for accurate work or a correction factor is applied. It should be pointed out that the *thermometers* used must be properly calibrated and that the *timer* used to measure the delivery time for the burets and pipets must also be calibrated.

Analytical Balances: The analytical balance (Figure 5.3), like volumetric glassware, is an example of a measuring device that cannot be adjusted to give a known response and, thus, if it does not give the known response (meaning if it does not correctly register a known weight), then it must be removed from service and repaired. The calibration procedure, then, consists of checking to see if it will register a known weight. If it does, within tolerable limits, it is calibrated and it may continue to be used. The "known weights" used for the calibration must be high-quality weights, each of which is certified as a known weight for calibration purposes ultimately by an organization such as NIST. The frequency of such calibration depends on the frequency of use, but it is not unusual to calibrate an analytical balance on a daily basis.

Standardization of a Titrant: For wet chemistry analytical methods, a titration is often used and the titrant, or the solution to which an unknown sample is compared, must be standardized. This can be done by comparing

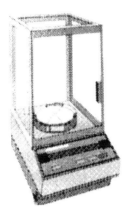

Figure 5.3 Analytical balance.

it with another standard. The important thing here is that the standard with which it is compared is ultimately traced to a SRM. The procedure utilizes volumetric glassware heavily, and thus the analyst must be assured that these are properly calibrated, as discussed above. *Auto-titrators* can be used (Figure 5.4). In this case, the automated equipment can be calibrated against manual equipment, i.e., volume readings obtained with the auto-titrator must match the volume readings obtained with a calibrated buret for the same sample. If they do not match (within accepted limits), the auto-titrator must be taken out of service and repaired, just like the defective balance.

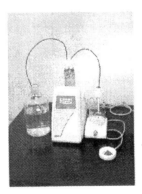

Figure 5.4 Auto-titrator.

Karl Fischer Titrators: These titrators measure moisture (water) in a variety of samples. The titrant's concentration is usually expressed as titer, or grams of water consumed per milliliter of titrant. Standardization involves a certified primary standard (a material containing a known amount of water). This standard is purchased in ampules and is accompanied by a test certificate indicating traceability to a reference material. In addition, the titrator should be calibrated for the titrant volume measurement. The moisture can be measued by weight loss upon drying and checked against the Karl Fischer results.

pH Meters: A pH meter (Figure 5.5) is calibrated (or "standardized") with the use of "buffer solutions," or solutions of known pH. A pH meter can be electronically adjusted to give the correct responses, the pHs of the buffer solutions. It can be calibrated using either one or two buffer solutions, but the pH values of these solutions should be in the range of the unknown solutions to be measured, since the meter would not be considered calibrated for pH values outside this range. A buffer solution can be certified as having a known pH by the vendor, but such certification must be traceable to SRMs certified by NIST.

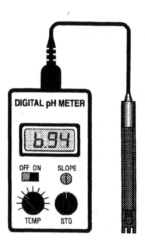

Figure 5.5   pH meter and probe.

Viscometers: Devices for measuring viscosity are called viscometers. The most common viscometer consists of a Cannon-Fenske tube, which is a U-shaped glass tube (see Figure 5.6), one arm of which consists of a capillary tube through which liquids flow slowly. The more viscous the liquid, the longer it takes for a given volume to flow through the capillary. This time is related to the viscosity of the liquid in poise or centipoise, which can be calculated from the measured time, a calibration constant, and the liquid's density. The viscosity is dependent on temperature, so the measurement must take place with the tube immersed in a constant-temperature bath. Calibration involves measuring this time for a standard substance, such as an oil for which the viscosity and density are known, in order to calculate

Figure 5.6   Cannon-Fenske viscometer. (Courtesy of Wilmad/LabGlass.)

the calibration constant. It is an example of a calibration for which the response for a standard substance must be established and then correlated with the unknowns.

Some viscosity tubes are calibrated at the factory, in which case a certificate of calibration, giving the calibration constant, is shipped with the tube. Such a constant can be checked in the laboratory using the method described above. Again, other measuring devices used in conjunction with this measurement must be properly calibrated. These include the temperature controller and the timer.

Spectrophotometers: For spectrophotometers, the absorbance responses for a series of solutions of an absorbing species at a particular wavelength of light are measured (see, for example, Figure 5.7). Beer's law, which governs

Figure 5.7 Spectronic Genesys 8 spectrophotometer. (Courtesy of Spectronic Corporation.)

spectrophotometric analyses, states that the absorbance is linear with concentration, and thus a "calibration curve," or "standard curve," a plot of absorbance vs. concentration, is expected to be linear, as in Figure 3.2 (Section 3). The responses of unknowns are then correlated with the known quantities via the standard curve to determine the concentration of analyte. Random errors occurring in the solution preparation and instrument functions usually result in points that do not fit a straight line exactly. However, a linear regression analysis can fit the best straight line to the plotted points, and a correlation coefficient, calculated from these data, indicates how well the absorbance values correlate with the concentration values. The linear regression analysis and correlation coefficient are typically done with a computer. In the normal scheme of things, a linear standard curve is thus taken to be an indication of a calibrated instrument. However, several potential variables require consideration, such as the standard solutions being traceable to a SRM, the cuvettes being matched in terms of pathlength and reflective/refractive properties, and the calibration of the wavelength control.

Atomic Absorption Spectrophotometers: This is a particular kind of spectrophotometer (see, for example, Figure 5.8) that utilizes a flame for the cuvette (requiring some maintenance for stability) and analyzes samples mostly for metals. The reference standards are thus solutions of metals. Such solutions are readily available and certified as being checked against NIST standards.

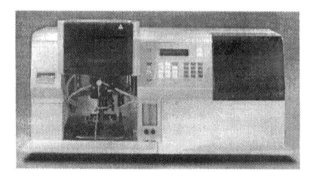

Figure 5.8   Perking-Elmer AA spectrophotometer. (Courtesy of Perkin-Elmer Corporation.)

Gas (GC) and Liquid (HPLC) Chromatographs: These are similar to spectrophotometers in that they are calibrated via a detector response to some property of analyte. The analyte may either be in solution or, in the case of GC (Figure 5.9), in pure form. Again, a calibration (or "standard") curve of detector response vs. either concentration or amount of pure chemical used is plotted and unknowns determined by correlation with the known standards via this curve. As with the spectrophotometer, the curve is linear. Variables here that require consideration for calibration include standards

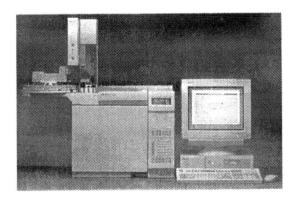

Figure 5.9   Hewlett-Packard gas chromatograph. (Courtesy of Hewlett-Packard.)

that are traceable to SRMs, quantity of standard measured via a syringe, and a stability of the detector response.

5.4   Reference Materials

In the above discussion, standard reference materials (SRMs) were mentioned often. A **reference material (RM)** is a material or substance suitable for use in calibrating equipment or standardizing solutions. A **certified reference material (CRM)** that a vendor indicates, via a certificate, is an RM. A **standard reference material (SRM)** is one that is distributed and certified by a certifying body, such as NIST. The SRM is the material to which all calibration and standardization materials should be traceable. A standard material becomes one when it is compared to or prepared from another. Ultimately, it all rests on the SRM — meaning all standard materials are traceable to an SRM (see Figure 5.10).

Figure 5.10   All calibration and standardization materials should be traceable to a SRM.

Reference materials are often real samples that have been carefully prepared and analyzed by many laboratories by many different methods. In this way, their known value and accompanying confidence limits are determined. The regular use of reference materials not only provides for calibration and standardization, but it also can demonstrate an analyst's proficiency with a method. It should be noted that SRMs are expensive, however, and are not often used for routine calibration and standardization work. Usually, primary and secondary standards are used for that. Another important fact about RMs is that they are considered to have a finite shelf life and cannot be confidently used as RMs after a certain period of time.

The concept of traceability is important to consider. As mentioned above, **traceability** is a standardization chain in which one material is established as a standard via a second standard, which was established as a standard via a third standard, etc. All secondary standards can be traced to a primary standard and this primary standard became a standard by comparison to an RM, ultimately being compared to an SRM.

Part of this chain is formed by the analyst in his/her laboratory (the "end user"), while part of it may be formed between NIST and the vendors. For example, a laboratory analyst can purchase a primary standard acid (which a vendor can certify as traceable to an SRM) for solution standardization and then base a number of secondary standardizations, such as acids and bases, on that one primary standard. Similarly, an analyst can purchase an atomic absorption reference standard (which a vendor can again certify as being traceable to an SRM) and then make one or more dilutions of this reference standard before creating the final series for the standard curve.

For his/her part, the laboratory analyst should be very concerned about maintaining the traceability chain. This is done by keeping good records and by providing complete labels for the containers. Good record-keeping includes the source and concentration of the material used, the identity and concentration of the standard being prepared, the name of the analyst who prepared it, the SOP used, the current date, and the expiration date if it is to be stored after the analysis is completed. A good label includes the standard's ID number (matching the notebook record), the name of the material and concentration, the date, the name of the analyst, and the expiration date. Record-keeping and labeling are addressed by ISO guidelines and by GLP regulations. Representative statements are presented in Box 5.4.

One final word about NIST is in order. The following statement found on the NIST Web site (www.nist.gov) provides some insight into the organization's role and mission: "The National Institute of Standards and Technology has pioneered and continues to be the leader in the development of certified reference materials for quality assurance of measurements through the Standard Reference Material Program, SRM. NIST provides more than 1300 different Standard Reference Materials (SRMs) that are certified for their special chemical or physical properties. SRMs are used

---

From ISO/IEC Guide 25
"(The laboratory) shall retain on record all original observations, calculations and derived data, calibration records and a copy of the calibration certificate, and test certificate or test report for an appropriate period. The records for each calibration and test shall contain sufficient information to permit their repetition."

From FDA Regulations 21 CFR 58 (GLP)
"Written records shall be maintained of all inspection, maintenance, testing, calibration, and/or standardizing operations."
"All reagents and solutions in the laboratory areas shall be labeled to indicate identity, titer or concentration, storage requirements, and expiration date."

Box 5.4   Statements from ISO guidelines and GLP regulations concerning record-keeping and labeling.

for three main purposes: (1) to help develop accurate methods of analysis (reference methods); (2) to calibrate measurement systems; and (3) to assure the long-term adequacy and the integrity of measurement quality assurance programs."

5.5   Statistical Control and Control Charts

Repeating a routine analysis over and over again for a period of time (perhaps sometimes years) and assembling the results into a data set that is free of bias and determinate errors create a basis for calculating a standard deviation that approaches σ, the true standard deviation. The $\pm 2\sigma$ theoretically associated with 95.5% of the values (Section 4.3), or the $\pm 3\sigma$ associated with 99.7% of the values then comes close to reality. If a given analysis result on a given day is then within $\pm 2\sigma$, it is a signal that "all is well" and the process or procedure is considered to be under what is called **statistical control**. If a process or procedure is under statistical control, then only 4.5% of the points (about 1 of every 20) would be outside the $\pm 2\sigma$ limits and only 0.3% (3 in 1000) would be outside the $\pm 3\sigma$ limits.

It is very important for any measurement process to be under statistical control in order to have some assurance that the results are reliable. An easy way to quickly see if a process or procedure is under statistical control is to maintain what is called a **control chart**. A control chart is a plot of a measurement or analysis result on the y-axis vs. time (usually days) on the x-axis. Included as horizontal lines are the $\pm 2\sigma$ limits and the $\pm 3\sigma$ limits. The $\pm 2\sigma$ limits are taken as **warning limits**, and the $\pm 3\sigma$ limits are taken as **action limits**. Any measurement or analysis result that falls between the warning limits and the action limits is cause for concern (1 out of every 20 occurs there naturally); and any that falls outside the action limits is cause for action, since only 3 out of every 1000 occur there naturally. An example of a control chart indicating that a process or measurement is under statistical control is given in Figure 5.11.

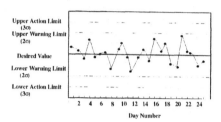

Figure 5.11 A control chart showing that a process is under statistical control because all the points are between the two warning limits.

As stated above, an occasional point outside the warning limits is expected. However, if there begins to be some consistency with points outside the warning limits, there is sufficient cause for some evaluation of the situation. Perhaps some component of the system is out of calibration, or perhaps some bias has been introduced inadvertently. Figure 5.12 is an example of such a control chart.

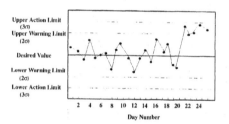

Figure 5.12 A control chart showing a consistent pattern outside the upper warning limit after Day 24, indicating that an evaluation of the situation is warrented. It may indicate a component of the system is out of calibration, etc.

If there is just one point outside the action limits, there should be an immediate evaluation. If there are two points or more in a brief period of time, the system should be shut down and analyzed for the cause. An example is Figure 5.13. As stated there, one possibility is a temporary contamination in the distilled water supply, which is certainly a concern even for a short period of time. A system out of statistical control, even for a short period of time, diminishes the reliability of any results.

A pattern occurring in control charts is probably caused by some discernable phenomenon. For example, if there is drift in the data either from low to high or high to low, it may be indicative of an electronic problem (see Figure 5.14). If there is a sudden shift in a pattern, it may be able to be traced to a sudden change that occurred in the analysis system,

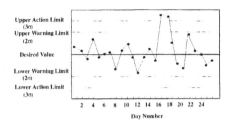

Figure 5.13 A control chart showing two consecutive days in which the measurement or analysis result was outside the action limit. An example of a cause is a distilled water line that was repaired, causing temporary contamination.

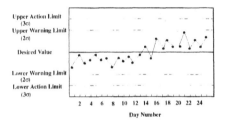

Figure 5.14 A control chart showing a steady change from low to high values. Such a control chart can result from a steady drift in the calibration of a spectrophotometric detector, for example, indicating that the detector may need to be repaired or replaced.

or in the material being sampled (see Figure 5.15). If there are sudden gyrations in the pattern from the upper action limit to the lower action limit, etc., it may be that something occurred that suddenly introduced new random errors, such as a new analyst on the job (see Figure 5.16). This kind of situation is seen as quite harmful to a laboratory's integrity and requires a careful study for answers.

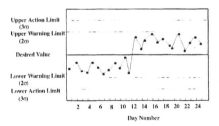

Figure 5.15 A control chart showing a sudden change, a "stair-step" change. This could result from a system component that was replaced, such as a light source in a spectrophotometer, or a recalibration of a component, such as a viscometer tube.

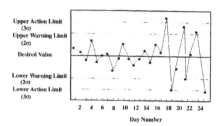

Figure 5.16 A control chart showing sporadic results starting with Day 18. This behavior would indicate the sudden introduction of random error, such as what might occur when a new analyst is assigned to the task.

Control charts are used in many laboratory situations in which careful daily monitoring would be beneficial to statistical control. Examples include not just analytical results, such as the specific gravity of a liquid pharmaceutical product, the level of a residual agricultural chemical in environmental waters, or the viscosity of a liquid polymer product; they are also useful in monitoring the calibration of various measuring devices in the laboratory rather than an analysis result. Examples include analytical balances (using known weights) and refractometers (using known liquids). Spectrophotometers (all types) and chromatographs (GC and HPLC) would utilize a **control sample**. A control sample, often referred to simply as a **control**, is a material of known composition (similar to the actual samples analyzed) used expressly for the purpose of monitoring a measurement process in the manner described here. The correct answer to the analysis is known because it was prepared like a standard and would be the "expected value" in a control chart.

5.6   *Method Selection and Development*

A variety of analytical methods may be available for a given analyte in a given system. For example, the analysis of drinking water for nitrate can be accomplished by any one of several methods, including UV and VIS spectrophotometry, ion-selective electrodes, ion chromatography, and capillary electrophoresis. A common activity in a quality assurance laboratory is to assess all available methods for a given analyte and then select the method that will best satisfy the needs. The analyst must determine what the needs are relative to such things as detection limit, bias, and accuracy, the useful range, capacity of the laboratory, cost, and ruggedness. The equipment and method must each be validated for the analysis in question, and this is followed by an evaluation of the entire system to assess its suitability for the project at hand.

Detection Limits: The detection limit is the smallest concentration value that an analyst can confidently say is detectable with the method being used. This value is one that must be distinguishable from the background electronic **noise** that is found for the **blank** for the analysis, which is a solution that contains all components of the sample except for the analyte. If the sensitivity of an electronic instrument, such as a spectrophotometer or chromatograph, is set on the highest level, the signal generated by the blank is electrically **noisy**, meaning it fluctuates in a continuous random manner within a range defined by a high level and a low level as the signal is being observed, such as on a computer screen. The fluctuations are due to random electrical impulses that occur naturally in any electronic device. Signals due to extremely small analyte concentrations can be "buried" in this noise and thus are not detectable.

The detection limit is often defined as the analyte concentration that produces a signal equal to twice the noise range, a **signal-to-noise ratio** of 2:1 (see Figure 5.17). While such an analyte concentration may be detectable (distinguishable from the noise), it does not mean that an accurate concentration can be determined. The signal due to the analyte should be well above the detection limit in order for the results to be reliable. A general rule of thumb is that the analyte signal should be at least five times noise level, a signal-to-noise ratio of 5:1, in order to be measured accurately.

Different methods have different detection limits. For example, the flame atomic absorption spectrophotometry (AAS) method for aluminum has a detection limit of 30 parts per million, while the inductively coupled plasma (ICP) method has a detection limit of 2 parts per million. Thus, if the sample solutions to be analyzed have a concentration level of less than 30 parts per million, then ICP must be used. On the other hand, if the concentration

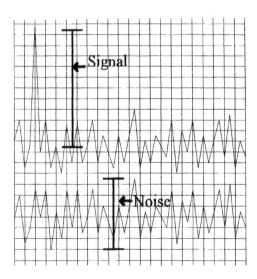

Figure 5.17 An illustration of signal and noise. Detection limit is defined as the concentration that produces a signal that is double the noise level.

levels are greater than 30 parts per million, then either method can be used. In that event, method selection would depend on other factors.

Bias and Accuracy: If there is a significant bias associated with a given method, it must be possible to compensate for it or the net results will not be accurate. Thus, while one may choose the ICP method mentioned above because of its lower detection limit, it may still not be a good choice because of an uncompensatable bias at the concentration level in question. It is appropriate to always check the method using a reference material, or by using an alternative method, to determine bias. Also, a bias can be caused by some correctable component of the analysis scheme, such as sampling problems or contamination, and not the method itself.

The Useful Range: The linear relationship evident in Figure 3.2 is not without limits. Almost all methods deviate from linearity at higher concentrations. It is important to determine at what concentration the linear range breaks down and begins to curve (Figure 5.18). It would not be accurate, in most cases, to determine a sample solution's concentration if its instrument response places it in a nonlinear range. One solution might be to dilute the sample solution in order to bring its concentration down to within the linear range and then compensate for the dilution mathematically with a dilution factor. It might also be possible to desensitize a method in some way in order to read the samples in a linear range. An example would be to use a shorter path length in spectrophotometry. Of course, the standards would also have to be read in this desensitized mode.

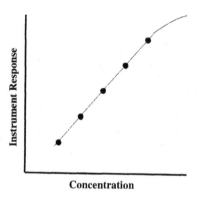

Concentration

Figure 5.18 Modification of Figure 3.2 showing a nonlinear range at the higher concentrations.

Capacity: Another consideration is the capacity of the laboratory to handle the work involved in a given method. For example, a capillary electrophoresis method would not be chosen if the laboratory does not have the instrument. It is also important to look at such factors as other equipment needed, the supplies needed, etc., or whether the laboratory can follow the required health and safety regulations, etc. Sometimes there may not be enough manpower or equipment to handle the sample work load. In that case, storage and refrigeration can also be a problem.

Cost: If the choice of method comes down to either Method A or Method B after having examined all facets of the work (detection limit, bias and accuracy, useful range, and capacity), the final decision may very well depend on cost. Cost includes all components of a budget (i.e., personnel, training, instrumentation, equipment, maintenance, supplies, overhead, and costs of using an outside laboratory for some or all of the work). All other factors being equal, if Method A is less expensive than Method B, then Method A should be chosen.

Ruggedness: Ruggedness refers to the attributes of a method that could cause the method to wear on the facilities, the staff, and the equipment. For example, there may be some aspect of the work that will cause an instrument component to deteriorate faster than it normally does. Perhaps one method requires 24-hour staffing, while an alternative method does not. Or perhaps the sample preparation task is too complicated to be done reliably.

Ruggedness also encompasses interferences, or sample components that produce a response in the measurement system that add to or subtract from the electronic signal due to the analyte. The analyst must be aware of such components in the particular sample system under investigation. Interferences can be controlled by removing them, masking them, or by correcting for their effects. A measurement system that is capable of measuring an analyte free of interferences even though other components are present is said to be **selective**. **Selectivity** can obviously be an important consideration in method selection.

Finding a Method: Various organizations publish volumes of methods for chemical analysis. One of the most well known is the **American Society for Testing and Materials**, or **ASTM**. The ASTM is a not-for-profit organization that provides a forum for producers, users, and consumers, to write standards for materials, products, systems, and services. The ASTM (Figure 5.19) publishes standard test methods encompassing metals, paints, plastics, textiles, petroleum, construction, energy, the environment, consumer products, medical services and devices, computerized systems, electronics, and many other areas. More than 10,000 ASTM standards are published each year in the 72 volumes of the *Annual Book of ASTM Standards*. Individual standards are also available.

Figure 5.19 The official logos of the United States Pharmacopeia (left), the American Society for Testing and Materials (center), and the AOAC International (right).

Another organization that publishes methods is **AOAC International** (Figure 5.19). AOAC stands for the Association of Official Analytical Chemists, which is the former name of this organization. AOAC International is an independent association of scientists in the public and private sectors devoted to promoting methods validation and quality measurements in the analytical sciences. The association's primary focus is coordination of the development and validation of chemical and microbiological analytical methods by expert scientists working in their industry, academic, and government laboratories worldwide. Methods proposed for their publication *Official Methods of Analysis of AOAC International* (*OMA*) are subjected to an eight-or-more laboratory collaborative study according to internationally recognized standards and receive rigorous scientific review of performance results. Once adopted they are published in the *Journal of AOAC International* and compiled for the OMA, which is periodically updated.

The **United States Pharmacopeia**, or **USP** (Figure 5.19), publishes methods for the pharmaceutical industry. The publication is called the *United States Pharmacopeia — National Formulary*, or *USP-NF*. It is published every 5 years and is a resource for drug standards and for ensuring the quality of drug products. It also provides standards for devices, diagnostics, and nutritional supplements. The mission of the USP is to promote public health by establishing and disseminating officially recognized standards of quality and authoritative information for the use of medicines and other health care technologies by health professionals, patients, and consumers.

Official methods of analysis are also published by the **Environmental Protection Agency (EPA)**, the **National Institute for Occupational Safety and Health (NIOSH)**, and the **Occupational Safety and Health Administration (OSHA)**.

Validation: **Validation** was defined in Section 3. It is the process of evaluating a method, an instrument or other piece of equipment, a standard material, etc. to determine whether it is appropriate for the work at hand and whether it will meet all expectations and needs for a given analysis. For example, an analyst may propose that a new gas chromatograph, one that has a new design of electron capture detector, be used for a certain pesticide analysis performed in the laboratory. A validation process would involve testing the new instrument (alongside the unit currently used in the procedure) with standards and samples used in the analysis to validate whether the new unit will perform up to the standards that have been set for the work. If it can be documented that the quality of the overall analysis by the new instrument meets expectations, then it can be brought "online."

An example of a piece of equipment that would need to be validated is a new gas chromatography column that a vendor is touting as especially useful for the work. For the validation study, the column is installed in the instrument and the procedure is executed, perhaps repeatedly on all types of possible samples, so that the analyst can be certain that, again, the quality of the work meets expectations.

The validation of a new method is more involved. A method may introduce a totally new system to the analysis, or it may introduce only new components to an established system. In any case, there may be several new techniques, several new pieces of equipment, several new standard materials, all of which need to be validated, both individually and as a unit. All of the method selection parameters mentioned above (detection limit, accuracy, etc.) are part of the validation process. Also, an important part of validation is the establishment of statistical control (see below).

System Suitability: System suitability refers to the validation of all components of an analysis system taken as a unit, a "system." For example, the analysis of an environmental water for pesticide residue involves a "method," which includes sampling (must represent the water in question), sample handling (e.g., what container is appropriate), sample preparation (perhaps an extraction process that includes the glassware, technique, timing, etc.), standards preparation (pipets, flasks, technique, etc.), injection technique, the instrument, and data handling (computer hardware and software), but it also includes individual pieces of equipment and instruments. The individual components of this system must be validated, but the system as a whole (all components working together) must also be evaluated in a system suitability evaluation scheme. The objective is the same: to determine whether the expectations, in terms of accuracy, precision, ruggedness, capacity, range, etc., are met.

Statistical Control for a New Method: To implement a new method, a laboratory must produce a preliminary track record of its success so that quality control charts can be established and then maintained. Aside from acquiring the space, supplies, equipment, instrumentation, and manpower required, the method must be tested, modified, tested again, etc., until it is ready to go "online." Gillis and Callio (listed in Bibliography) recommend the following sequence for preparing an instrumental method for routine use.

1. A first trial using a series of blanks (4 to 5) and a series of standards (10 to 15). Distilled water (or pure solvent) is used for instrument zeroing. This gives the analyst the opportunity to get a feel for the method, to identify the records that must be kept, to identify any snags or bottlenecks, and to make changes.
2. After making changes, a second trial run using the same blanks and standards. Again, any desired changes are made.
3. Additional trial runs until the analyst is confident that he/she can move on without making any further changes.
4. Another series of trials, all identical to each other (no changes). This time, the results should be tabulated, and a mean and a standard deviation for the blank and each standard should be calculated and the data graphed (mean response values vs. concentration) to create the standard curve (Figure 3.2). In addition, the slope of the line and the y-intercept are determined, as well as the correlation coefficient.

Figure 5.20 Implementing a new method involves acquiring and examining a large volume of data.

If the results look good, one moves on to Step 5, or makes some change to try to improve the results and repeat the above process.

5. First Control Run. A large number (7 to 15) of sets of standards and blanks are run and the results tabulated, as in the trial runs. These data are then plotted (responses vs. concentration for all data points, on one graph) and the means, standard deviations, RSDs, the slope, y-intercept, and correlation coefficient are determined. The smaller the value of the y-intercept, the better (the less chance for a contamination or interference problem). The closer the slope is to 1, the better (the more sensitive). At higher concentrations, the standard deviation should get larger, and the RSDs should get smaller (while approaching some limit). If the RSDs are between 30% and 100%, a close approach to the detection limit is indicated.

6. Second Control Run. The second run is compared to the first. Improved slope, smaller blank values, and smaller standard deviations are desirable. More control runs are done if necessary to improve performance. Real samples are run when the performance is consistent.

7. Real samples. The move to analyze real samples represents a move toward the unknown. Not only are the results of the analysis unknown ahead of time, but other variables relating to sample inhomogeneity, sample preparation variables, additional sources of error, etc. are introduced. A large number (>30) of duplicate samples should be analyzed so that a reliable standard deviation and a reliable control chart can be established. The ultimate purpose of this work is to characterize what is a typical analysis for this kind of sample so that one can know when the method is under statistical control and when it is not. Charting the responses of a standard, the slope of the standard curve, the blanks, spike recoveries, etc. is important.

5.7 Proficiency Testing

Proficiency testing is an activity in which the competence of the workers in a laboratory, or set of laboratories, is determined. This is done by comparing the results obtained for a sample or reference material with either the results of the other laboratories (real sample) or with the true answer (reference material), or both. A study coordinator manages the activity, reports the results, and grades the participating laboratories (Figure 5.21). This can be crucial for a given laboratory because its reputation is at stake and contractual work may hang in the balance. Of course, a confident laboratory may view it in a positive light because it will prove that they are competent and proof of their proficiency "would look good on their resume."

Sometimes, when evaluating the results of one lab compared to another, one analyst compared to another, or one instrument compared to another, a serious variability is observed. In other words, different answers to an analysis, or different individual data points, are obtained for no apparent reason and seem to depend on what lab does the work, or what analyst does the work, or which instrument is used. The determination of the individual biases that can occur can be frustrating to the lab managers. The key to this dilemma, although sometimes inadequate, is to minimize the variables as much as possible.

Figure 5.21 A good grade on a proficiency test is a "gold star" for a laboratory.

6 The sample

In the preceding discussions, the "sample" has been mentioned more than just a few times. The composition of the sample is absolutely critical to a successful and accurate analysis. Presented below is some essential information regarding this important factor of analytical science.

6.1 Definitions and Examples

A **sample** is a small portion of a larger body of material that is obtained for laboratory analysis. If the purpose of such analysis is to present the concentration of a component in the entire body of material, as it usually always is, then this sample must be representative of the entire body. A **representative sample** is therefore a sample that possesses all the characteristics of a larger bulk system in exactly the same concentration levels as in the system. In other words, it represents the system and whatever concentration level is found for a given component of a sample, that is also then taken to be the concentration level in the system. For example, in order to analyze the water in a lake for mercury, a bottle is filled with the water and then taken into the laboratory for analysis. If the mercury level in this sample is determined to be 12 ppm, then assuming the sample is representative of the entire lake, the entire lake is assumed to be 12 ppm in mercury, also.

Sometimes, in order to better assure the "representative" status, a series of samples are obtained from different parts of the bulk system and then combined into one sample. This kind of sample is called a **composite sample**. A composite sample is useful in situations in which it is likely that the substance to be determined is not homogeneously distributed throughout the entire system. For example, when determining the fertilizer needs of a person's lawn, it might be best to obtain a composite sample because the phosphate level (for example) in one part of the lawn might not be the same as in another part.

Another method for solving the problem of non-homogeneity of a bulk system is to take a **selective sample**. A selective sample is a sample that is obtained from a particular part of the bulk system that is known, or assumed, to have a different composition. For example, the air next to a leaking gas furnace exhaust would have a higher level of carbon monoxide than the air in a room elsewhere in a building. In the case in which the carbon monoxide level in the immediate vicinity of the leak is important, then a selective sample is acquired.

If it is thought that a bulk system is homogeneous for a particular component, then a **random sample** is taken. This would be just one sample taken from one location at random in the bulk system. An example would be when determining the level of active ingredient in a pharmaceutical product stored in boxes of individual bottles in a warehouse. Having no reason to assume a greater concentration level in one bottle or box compared to another, a sample is chosen at random.

Other designations for samples are **bulk sample, primary sample, secondary sample, subsample, laboratory sample,** and **test sample.** These terms are used when a sample of a bulk system is divided, possibly a number of times, before actually being used in an analysis. For example, a water sample from a well may be collected in a large bottle (bulk sample or primary sample), from which a smaller sample is acquired by pouring into a vial to be taken into the laboratory (secondary sample, subsample, or laboratory sample), then poured into a beaker (another secondary sample or subsample), before a portion is finally carefully measured into a flask (test sample) and diluted to make the sample solution.

How the sample is obtained, how it is handled, how it is prepared, etc. are obviously, then, very important parts of the work of an analytical laboratory. The results of an analysis can only be useful if the sample really does represent what it is intended to represent. The axiom "an analysis is only as good as the sample" is absolutely true.

6.2   *Statistics of Sampling*

A consideration of statistics is required in a discussion of sampling because of the randomness with which samples are acquired. We may take pains to see that a sample randomly acquired is representative, but the compositions of a series of such samples taken from the same system vary according to the location of the sampling and other factors. So the sampling is similar to a laboratory analysis — results vary randomly and are affected by random errors that cannot be compensated — thus, the need for statistics, as discussed in Section 4.

A novice might think that a lab analyst obtains a single sample from a bulk system, analyzes it one time in the laboratory, and reports the answer to this one analysis as the analysis results. If variances in the sampling and lab work are both insignificant, these results may be valid. However, due to possible large variances in both the sampling and the lab work, such a result cannot be considered reliable. The discussion in Section 4 indicated that the correct procedure is to perform the analysis many times and deal with the variances with statistics.

The fact that sampling introduces a second statistics problem means that one must also consider taking a large number of samples and deal with the results with statistics just as one performs a laboratory analysis a large number of times and deals with those results with statistics. For example, it can be shown that if a measurement system generates data with a standard deviation of 10 ppm and one needs to know an average concentration to ±5 ppm with 95% confidence, then one must perform the analysis 16 times. If the sampling variance is high but the lab analysis variance is low, then one must measure 16 samples each one time. If the lab analysis variance is high but the sampling variance is low, then one must measure one sample 16 times. If both the sampling variance and the lab analysis variance are high, then one must measure 16 samples each 16 times.

Chemists want to have as low a variance (or standard deviation) as possible for the greatest accuracy. If it is not possible to have a low enough standard deviation to suit the need, then the number of measurements (either the number of samples, the number of lab analyses, or both) must be increased. If increasing the number of measurements is not desirable (due to an increased workload or expense, for example), then the chemist must live with a larger error.

6.3   Sample Handling

The importance of a high-quality representative sample has already been noted. How to obtain the sample and what to do with it once it reaches the laboratory are obviously important factors. But the handling of the sample between the sampling site and the laboratory is often something that is given less than adequate consideration (Figure 6.1). The key concept is that the

Figure 6.1   Handling the sample between the sampling site and the lab is often given less than adequate consideration.

sample's integrity must be strictly maintained and preserved. If a preserva-
tive needs to be added, then someone needs to be sure to add it. If maintaining integrity means to refrigerate the sample, then it should be adequately refrigerated. If there is a specified holding time, then this should be
accurately documented. If preventing contamination means a particular
material for the storage container (e.g., glass vs. plastic), then that material
should be used.

It is very important to document who has handled the sample and what
responsibility each handler has at various junctures between the sampling
site and the laboratory. In other words, the **chain of custody** must be
maintained and documented. A sample can have a number of custodians
along the way to the laboratory. A sample of lake water may be taken by a
sampling technician at the site. The sampling technician may give it to a
driver who transports it to the analysis site. A shipping/receiving clerk may
log in the sample and give it to a subordinate who takes it to the laboratory.
Along the way, this sample is in the hands of five different handlers — the
sampling technician, the driver, the shipping/receiving clerk, the subordinate, and the laboratory technician. Each should maintain documentation
of his/her activity and duties and copies of the chain of custody should be
filed (see Figure 6.2).

Figure 6.2   Documentation of the chain of custody is important for sample integrity.

7 Good laboratory practices

In Section 2, it was noted that various laboratories are under federal regulation in relation to the activities and the quality of the work performed in the laboratory. These regulations have come to be known as Good Laboratory Practices, or GLP. There are two sets of GLP regulations: one for the Food and Drug Administration (FDA) and one for the Environmental Protection Agency (EPA). They differ only in statements relating to these agencies' objectives and purposes. The major subparts of the GLP regulations are listed in Box 7.1. The following subsections take a close look at these regulations.

7.1 General Provisions

The first section in each set is titled "General Provisions." In this section, in separate numbered paragraphs, the scope of the regulations is laid out (see Boxes 7.2 and 7.3), a number of definitions are listed, and the applicability of the regulations to studies performed under grants and contracts is covered. Also presented is a paragraph that indicates that the inspection of a testing facility is permitted. Besides these, the EPA GLP includes a statement of compliance or non-compliance as well as a statement on the effects of non-compliance.

Outline of GLP Regulations

Subpart A: General Provisions
Subpart B: Organization and Personnel
Subpart C: Facilities
Subpart D: Equipment
Subpart E: Testing Facilities Operation
Subpart F: Test, Control, and Reference Substances
Subpart G: Protocol for and Conduct of a Study
Subpart H, I: Reserved
Subpart J: Records and Reports
Subpart K (FDA only): Disqualification of Testing Facilities

Box 7.1   Major subparts of the GLP regulations, both EPA and FDA.

FDA
21 CFR 58

Part 160 - GOOD LABORATORY PRACTICES FOR NONCLINICAL LABORATORY STUDIES STANDARDS

Subpart A - General Provisions

§58.1 Scope

(a) This part prescribes good laboratory practices for conducting nonclinical laboratory studies that support or are intended to support applications of research or marketing permits for products regulated by the Food and Drug Administration, including food and color additives, animal food additives, human and animal drugs, medical devices for human use, biological products, and electronic products. Compliance with this part is intended to assure the quality and integrity of safety data filed pursuant to sections 406, 408, 409, 502, 503, 505, 506, 507, 510, 512–516, 518–520, 706 and 801 of the Federal Food, Drug, and Cosmetic Act and sections 351 and 354–360F of the Public Health Service Act.

(b) References in this part to regulatory sections of the Code of Federal Regulations are to Chapter 1 of Title 21, unless otherwise noted.

Box 7.2   The beginning of the Code of Federal Regulations, Chapter 21, Part 58, and the statements of defining the scope of the FDA GLP.

EPA
40 CFR 160

Part 160 - GOOD LABORATORY PRACTICE STANDARDS

Subpart A - General Provisions

§160.1 Scope

(a) This part prescribes good laboratory practices for conducting studies that support or are intended to support applications of research or marketing permits for pesticide products regulated by the EPA. This part is intended to assure the quality and integrity of data submitted pursuant to sections 3, 4, 5, 8, 18 and 24(c) of the Federal Insecticide, Fungicide, and Rodenticide Act (FIFRA), as amended (7 U.S.C. 136a, 136c, 136f, 136q, and 136v(c) and sections 408 and 409 of the Federal Food, Drug and Cosmetic Act (FFDCA) (21 U.S.C. 346a, 348).

(b) This part applies to any study described by paragraph (a) of this section which any person conducts, initiates, or supports on or after October 16, 1989.

Box 7.3   The beginning of the Code of Federal Regulations, Chapter 40, Part 160, and the statements defining the scope of the EPA GLP.

The scope paragraph of the EPA GLP mentions that the research and marketing of pesticide products are regulated. In addition, all data relating to certain sections of FIFRA, the Federal Insecticide, Fungicide, and Rodenticide Act, are regulated. The scope paragraph of the FDA GLP mentions that the research and marketing of products regulated by the FDA are regulated. Specific products are listed. In addition, under FDA GLP, all data relating to certain sections of the Federal Food, Drug, and Cosmetic Act (FFDCA) and the Public Health Service Act (PHSA) are regulated.

The definitions are for those words and phrases that the reader encounters in the regulations. Examples include quality assurance unit, raw data, reference substance, sponsor, study, study director, test substance, test system, testing facility. See Box 7.4 for these definitions.

In the paragraph entitled "Applicability to studies performed under grants and contracts," the regulations state that whenever a sponsor utilizes the services of a consulting laboratory, contractor, or grantee to perform an analysis or other service, it shall notify this laboratory, contractor, or grantee that the service must be conducted in compliance with GLP. Hence, essentially all phases of the work are covered by the regulations, whether performed by an outside organization or the sponsor's own laboratory.

The paragraph that discusses the "Inspection of a testing facility" is of significant importance. It is found in both the EPA and FDA GLP and, in essence, allows government representatives to enter the facilities and inspect and copy all records within the scope of the work at reasonable times and in a reasonable manner. The text of the paragraph found in the EPA regulations is reproduced in Box 7.5. A key statement here is that the EPA (or FDA) will not consider a study in support of an application for a permit if the testing facility refuses to permit inspection. Thus a facility must permit inspection. This inspection, or "audit" as it is often called, is therefore taken very seriously. See Section 8 for more information.

The EPA regulations include a statement of compliance or non-compliance (§160.12) and a statement indicating the effects of non-compliance (§160.17), while the FDA regulations address these issues in their Subpart K. The EPA statements say that any person who submits an application for a research or marketing permit shall submit a statement signed by the applicant, the sponsor, and the study director that the study was either conducted in accordance with GLP, conducted in part in accordance with GLP (with those parts not so conducted described), or conducted such that it was not known whether it was in accordance with GLP. As to non-compliance consequences, the EPA regulations say that the EPA may refuse to consider reliable any data from a study which was not conducted in accordance with GLP and that severe penalties may result from submission of a compliance statement that is false.

The FDA statements in Subpart K are much more explicit and detailed. While the EPA statements are described in just the two paragraphs mentioned above, the FDA statements are written in a total of eight paragraphs. The paragraph numbers and titles are listed in Box 7.6.

Quality Assurance Unit: Any person or organizational element, except the study director, designated by testing facility management to perform the duties relating to quality assurance of the (nonclinical laboratory) studies.

Raw data: Any laboratory worksheets, records, memoranda, notes, or exact copies thereof, that are the result of original observations and activities of a (nonclinical laboratory) study and are necessary for the reconstruction and evaluation of the report of that study.

Reference Substance (EPA only): Any chemical substance or mixture, or analytical standard, or material other than a test substance, feed, or water, that is administered to or used in analyzing the test system in the course of a study for the purposes of establishing a basis for comparision with the test substance for known chemical or biological measurements.

Person: An individual, partnership, corporation, association, scientific or academic establishment, government agency, or organizational unit thereof, and any other legal entity.

Sponsor: (1) A person who initiates and supports, by provision of financial or other resources, a (nonclinical laboratory) study; (2) a person who submits a (nonclinical laboratory) study to the EPA (FDA) in support of an application for a research or marketing permit; or (3) a testing facility, if it both initiates and actually conducts the study.

Study (EPA): An experiment in which a test substance is studied in a test system under laboratory conditions or in the environment to determine or help predict its effects.

Nonclincial Laboratory Study (FDA): An experiment in which test articles are studied prospectively in test systems under laboratory conditions to determine their safety.

Study Director: The individual responsible for the overall conduct of a (nonclinical laboratory) study.

Test Substance (EPA): A substance or mixture administered or added to a test system in a study, in which the substance or mixture (1) is the subject, or contemplated subject, of an application for a research or marketing permit, or (2) is an ingredient or product of a substance as defined above.

Test Article (FDA): Any food additive, color additive, drug, biological product, electronic product, medical device for human use, or any other article subject to regulation under the FFDCA or PHSA.

Test System (EPA): Any animal, plant, microorganism, chemical or physical matrix, including but not limited to soil or water, or subparts thereof, to which the test, control, or reference substance is administered or added for the study.

Test System (FDA): Any animal, plant, microorganism, or subparts thereof, to which the test or control article is administered or added for study.

Testing Facility (both EPA and FDA): A person who actually conducts a (nonclinical laboratory) study, i.e., who actually uses the test substance (article) in a test sytem.

Box 7.4   Definitions of some terms selected from the EPA and FDA GLP regulations. The words "nonclinical laboratory" precede the word "study" in the FDA definitions.

§ 160.15 Inspection of a Testing Facility
(a) A testing facility shall permit an authorized employee or duly designated representative of EPA or FDA, at reasonable times and in reasonable manner, to inspect the facility and to inspect (and in the case of records also to copy) all records and specimens required to be maintained regarding studies to which this part applies. The records inspection and copying requirements should not apply to quality assurance unit records of findings and problems, or to actions recommended and taken, except that EPA may seek production of these records in litigation or formal adjudicatory hearings.
(b) EPA will not consider reliable for purposes of supporting an application for a research or marketing permit any data developed by a testing facility or sponsor that refuses to permit inspection in accordance with this part. The determination that a study will not be considered in support of an application for a research or marketing permit does not, however, relieve the applicant for such a permit or any obligation under any applicable statute or regulation to submit the results of the study to the EPA.

Box 7.5 The EPA GLP statement concerning the Inspection of a Testing Facility. The FDA statement is very similar.

Subpart K Paragraphs, FDA GLP Regulations
§ 58.200 Disqualification of Testing Facilities
§ 58.202 Grounds for Disqualification
§ 58.204 Notice of and Opportunity for Hearing on Proposed Disqualification
§ 58.206 Final Order on Disqualification
§ 58.210 Actions Upon Disqualification
§ 58.213 Public Disclosure of Information Regarding Disqualification
§ 58.215 Alternatives or Additional Actions to Disqualification
§ 58.217 Suspension or Termination of a Testing Facility by a Sponsor
§ 58.219 Reinstatement of a Disqualified Testing Facility

Box 7.6 Paragraphs within the FDA GLP regulations relating to compliance or non-compliance with the regulations.

7.2 Organization and Personnel

Sections with this title appear under both the FDA and EPA regulations and are essentially identical. Subsections include requirements for personnel (FDA §58.29 and EPA §160.29), requirements for testing facility management (FDA §58.31 and EPA §160.31), requirements for the study director (FDA §58.33 and EPA §160.33), and requirements for the quality assurance unit (FDA §58.35 and EPA §160.35).

The subsections on personnel state that each individual engaged in the work must have the proper education, training, and/or experience; that each testing facility must maintain a file summarizing the education, training, and experience of the workers; and that there shall be a sufficient number of workers to complete the study in a timely manner. In addition, these subsections state that the workers follow sanitation and health procedures; that the workers must wear appropriate clothing and that this clothing be changed as necessary to prevent contamination; and that any worker found to be ill be appropriately excluded from contact with any operation or function that may adversely affect the study.

The subsections on testing facility management specifically state the duties of the management. These include (a) designating the study director before the study is initiated, (b) replacing the study director if necessary, (c) assuring that there is a quality assurance unit, (d) assuring that all test substances (articles) involved in the study be tested for identity, strength, purity, stability, and uniformity, as applicable, (e) assuring that personnel, resources, facilities, equipment, materials, and methodologies are available as scheduled, (f) assuring that personnel clearly understand their functions, and (g) assuring that any reported deviations from regulations are communicated to the study director and corrective actions are taken and documented.

The subsections outlining the requirements of the study director, the individual responsible for the overall conduct of the study, state that he/she be a scientist or other professional of appropriate education, training, and experience (Figure 7.1). Additionally, this individual is responsible for the interpretation, analysis, documentation, and reporting of results, and represents the single point of study control. He/she assures that (a) the protocol (see Section 7.7), including any changes, is approved and followed, (b) all experimental data are accurately recorded and verified, (c) unforeseen circumstances affecting the quality and integrity of the study are noted and corrective action taken and documented, (d) test systems are identified in the protocol, (e) all applicable GLP are followed, and (f) all raw data,

Figure 7.1 The study director must be a professional of appropriate education, training, and experience.

documentation, protocols, specimens, and final reports are transferred to the archives during or at the close of the study.

Finally, the subsections outlining the requirements of the quality assurance unit state the fact that such a unit is required; state the duties of this unit; state that the specific characteristics of the unit, including responsibilities, applicable procedures, and records, must be maintained in writing; and state that the EPA or FDA shall have access to the documents. The quality assurance unit (see definition in Box 7.4) is responsible for monitoring each study to assure management that the facilities, equipment, personnel, methods, practices, records, and controls are in conformance with the regulations. The quality assurance unit must be completely separate from and independent of the personnel engaged in the direction and conduct of the study. A listing of the specific duties of the quality assurance unit is presented in Box 7.7.

Duties of the Quality Assurance Unit
1. Maintain a copy of a master schedule sheet of all (nonclinical laboratory) studies conducted at the testing facility.
2. Maintain copies of all protocols pertaining to all (nonclinical laboratory) studies for which the unit is responsible.
3. Inspect each (nonclinical laboratory) study, maintain written records of these inspections providing the details of each, and report any problems to the study director.
4. Submit periodic written reports to the study director and to management.
5. Determine that no deviation occurred without proper authorization or documentation.
6. Review the final study report for accuracy.
7. Include an official signed statement of the inspections with the final report.

Box 7.7 The duties of the quality assurance unit as specified in FDA §58.35 and EPA §160.35.

7.3 Facilities

The facilities for the study are described in Subpart C for both the EPA and the FDA regulations. There are six subsections (or paragraphs) under Subpart C in each case and the titles of these subsections are very similar (see Box 7.8). The key points here are (1) facilities must be of suitable size and construction and must be designed so that they are separate from other activities so as to prohibit adverse effects on the study from the other activities; (2) proper separation of test systems, including waste disposal systems, must be maintained; (3) storage areas for feed, nutrients, soils, bedding, supplies, and equipment shall be provided and the areas for storing feed, nutrients, soils, and bedding shall be separated from the test systems, and, in addition, these facilities shall be provided as required by the written protocol; (4) facilities for handling test articles, control articles, and reference substances, including receipt and storage, mixing, and storage of prepared

---

Subpart C - Facilities

FDA	EPA
§ 58.41 General	§ 160.41 General
§ 58.43 Animal care facilities	§ 160.43 Test system care facilities
§ 58.45 Animal supply facilities	§ 160.45 Test system supply facilities
§ 58.47 Facilities for handling test and control articles	§ 160.47 Facilities for handling test, control, and reference substances
§ 58.49 Laboratory operation areas	§ 160.49 Laboratory operations area
§ 58.51 Specimen and data storage facilities	§ 160.51 Specimen and data storage facilities

---

Box 7.8   The subsections under Subpart C of the FDA and EPA GLP regulations.

mixtures, shall be organized so as to prevent mixups and to preserve the identity, strength, purity, and stability of the articles and mixtures; (5) special laboratory space shall be provided as needed for all procedures; and (6) there shall be archives, limited to access by authorized personnel, for storage of study records and specimens.

7.4   Equipment

Subpart D presents the regulations for the equipment used in a study. There are two subsections in both sets, the FDA and the EPA GLP: one for equipment design and one for maintenance and calibration. The text of both sets are virtually identical, stating that all equipment must perform as required by the protocol and that this equipment must be maintained and calibrated. Formal SOPs must be on file for this and must set forth in sufficient detail the methods, materials, and schedules to be used in the inspection, cleaning, maintenance, testing, calibration, and/or standardization of equipment and also name the person responsible. In addition, written records shall be maintained for this activity, including records of any malfunctions, how these were discovered, and what corrective action was taken. This subpart was mentioned previously in the discussion of calibration of equipment (Section 5.3).

7.5   Testing Facility Operations

Subpart E of both the FDA and the EPA GLP regulations addresses the operations aspects of the work. This is the subpart that specifies the use and design of standard operating procedures that was discussed in Section 5.2 (see Box 5.1 in which the text of EPA Subsection 160.81 is reproduced). This is also the subpart that specifies the labeling of reagents and solutions that was mentioned in Section 5.4 (see Box 5.4 in which the FDA Subsection 58.83 is reproduced). Some important points in the subsection describing standard operating procedures are (1) SOPs that satisfy management of the quality and integrity of the data are required; (2) the study director may authorize deviations from an SOP

Figure 7.2   All deviations from SOPs are authorized by the study director.

(Figure 7.2); (3) significant changes must be authorized by management; (4) SOPs must be immediately available to personnel; (5) an historical file of SOPs must be maintained. An important point regarding labeling is that *every* bottle in the laboratory must have a label with the required information. The required information, as delineated in this subpart, includes identity, titer or concentration, storage requirements, and expiration date.

There is also a subsection in Subpart E providing considerable detail as to the care of test systems that are animals or plants.

7.6   *Test, Control, and Reference Substances*

Test substances (or "articles"), control substances (or "articles"), and reference substances are covered by Subpart F in both the FDA and EPA regulations. Test substances (EPA) and test articles (FDA), as well as reference substances, were defined in Box 7.4, Section 7.1. Basically, this subpart covers all substances under investigation and all known substances used in the investigation in terms of their characterization, handling, and mixing.

First, in terms of their "characterization," the regulations state that the identity, strength, and other characteristics shall be determined and documented before use. The regulations also state that their stability shall be determined before the experimental start date, or concurrently with the study. In addition, test substance storage containers must be labeled and must be kept for the duration of the study and, for studies lasting more than 4 weeks, reserved samples must be retained according to Subsections 58.195

(FDA) and 160.195 (EPA). Finally, the EPA (but not the FDA) regulations state that the stability of the substances under storage conditions at the test site shall be known for all studies.

Second, in terms of handling, the regulations state that procedures must be established to ensure that (a) there is proper storage; (b) contamination or deterioration is avoided during handling; (c) proper identification is maintained throughout the study; and (d) the receipt and distribution is documented for each batch. The concept of the chain of custody documentation (Section 6.3) is thus covered in the regulations.

Third, for each substance (or "article") that is mixed with a carrier (solvent or other medium), (1) the uniformity or concentration shall be determined; (2) the solubility shall be determined (EPA only); and (3) the stability in the mixture shall be determined. Expiration dates shall be clearly shown and (EPA only) the device used to formulate the mixture shall not interfere with the integrity of the test.

7.7   Protocols for and Conduct of a (Nonclinical Laboratory) Study

The term "protocol" has a specific meaning in GLP. It is defined in Subpart G of both the FDA and EPA GLP as an official written document that clearly indicates the objectives and all methods for the conduct of the study. In the case of the EPA GLP, it contains a set of 15 specific items. In the case of the FDA GLP, it contains a set of 12 specific items. These are summarized in Box 7.9. Like SOPs, an approved protocol can be changed or revised, but the changes and revisions must be documented, signed by the study director, dated, and maintained with the original document.

A specific path for conducting a study is also outlined in Subpart G. Besides stating that the study shall be conducted and test systems monitored in accordance with protocol (Figure 7.3), and that specimens shall be properly identified

Figure 7.3   Studies must be conducted according to the official written protocol.

List of Required Items for a Protocol of a (Nonclinical Laboratory) Study (From EPA §160.120 and FDA §58.120)

Both the EPA and FDA GLP:
1) A descriptive title and statement of purpose of the study.
2) Identification of the test, control, and reference substances by name, chemical abstracts service (CAS) number or code number.
3) The name and address of the sponsor and the name and address of the testing facility at which the study is being conducted.

EPA GLP only:
4) The proposed experimental start and termination dates.
5) Justification for the selection of the test system.

Both EPA GLP and FDA GLP:

4 & 6)	The number, body weight range, sex, source of supply, species, strain, substrain, and age of the test system.
5 & 7)	The procedure for the identification of the test systems.
6 & 8)	A description of the experimental design, including methods for the control of bias.
7 & 9)	Where applicable, a description and/or identification of the diet used in the study as well as solvents, emulsifiers and/or other materials used to solubilize or suspend the test, control, or reference substances before mixing with the carrier. The description shall include specifications for acceptable levels of contaminants that are reasonably expected to be present in the dietary materials and are known to be capable of interfering with the purpose or conduct of the study if present at levels greater than established by the specifications.

EPA GLP only:
10) The route of administration and the reason for its choice.

Both EPA and FDA GLP:

8 & 11)	Each dosage level, expressed in milligrams per kilogram of body or test system weight or other appropriate units, of the test, control, or reference substance to be administered and the method and frequency of administration.
9 & 12)	The type and frequency of tests, analyses, and measurements to be made.
10 & 13)	The records to be maintained.
11 & 14)	The date of approval of the protocol by the sponsor and the dated signature of the study director.
12 & 15)	A statement of the proposed statistical methods to be used.

Box 7.9   The required items for a protocol as delineated in EPA §160.120 and FDA §58.120.

and results made available to a pathologist, an important statement regarding the data generated is presented. This statement is reproduced in Box 7.10.

The EPA GLP includes a subsection under Subpart G (§160.135) that deals with physical and chemical characterization studies. This subsection states that all provisions of GLP standards apply to certain specific physical and chemical characterization studies of test, control, and reference substances. These studies are listed. It also states that certain specified subsections do not apply to studies other than those listed.

7.8   Records and Reports

Subpart J in both the FDA and EPA GLP deals with the records and reports generated by a study. Specifically, this subpart, for both the EPA and the FDA, deals with the reporting of the study results, the storage and retrieval of records and data, and the retention of records. Under "report of the study results" (EPA §160.185 and FDA §58.185), the items to be included in the final report are listed. The list includes 14 enumerated items, including the following: names, dates, objectives and procedures, statistical methods, identities of and data concerning substances used, laboratory methods, test systems, dosages, data integrity issues, specific data handling procedures, reports of scientists involved, data storage locations, and the quality assurance unit statement. Corrections or additions to the final report are handled via amendments.

All raw data, documentation, records, protocols, specimens, final reports, and correspondence relating to data interpretation must be retained and archived according to the "storage and retrieval of records and data" subsections (EPA §160.190 and FDA §58.190) found in this subpart, and these records must be available for expedient retrieval. The archived records shall be carefully protected and indexed.

Statement Concerning Data Generated While Conducting a Study
(EPA GLP § 160.130(e) and FDA GLP §58.130)

All data generated during the conduct of a (nonclinical laboratory) study, except those that are generated by automated data collection systems, shall be recorded directly, promptly, and legibly in ink. All data entries shall be dated on the date of entry and signed or initialed by the person entering the data. Any change in entries shall be made so as not to obscure the original entry, shall indicate the reason for such change, and shall be dated and signed, or identified at the time of change. In automated data collection systems, the individual responsible for the direct data input shall be identified at the time of data input. Any change in automated data entries shall be made so as not to obscure the original entry, shall indicate the reason for change, shall be dated, and the responsible individual shall be identified.

Box 7.10   Statement concerning data generated while conducting a study as presented in the indicated GLP sections.

Additional statements regarding records retention, which do not supersede the previous statements, are presented in the "retention of records" subsection (EPA §160.195 and FDA §58.195). The period of time specific records must be retained is indicated here.

8 Audits

Laboratories may indicate that they do quality work and adhere to GLP regulations and/or ISO guidelines, but they must also prove that they do so. Thus, the GLP regulations provide for audits by the quality assurance unit as well as by EPA or FDA inspectors, and ISO registration will also involve an audit process.

8.1 Quality Assurance Audit

The quality assurance unit (QAU) performs the quality assurance audit. This task is ongoing throughout the study and is the reason for the existence of the QAU. The QAU is the person (see definition of "person" in Box 7.4) designated to perform the quality assurance duties. Thus, the QAU inspects the study to assure integrity and that problems are brought to the attention of the study director. This unit also determines if unauthorized deviations from protocols and SOPs occurred and reviews the final report for accuracy.

The parts of the study that are included in the inspection include the facility, the manner in which the study is being conducted, the data generated, the protocol, and the final report. If any of the work is "farmed out" to a contract lab, or if any of the work otherwise occurs off-site, then these same items at the contract lab or at the field site also come under scrutiny for the QAU.

Identity of the QAU: As stated in Section 7.2, the QAU must be completely separate from and independent of the personnel engaged in the direction of and conduct of the study. While it can be an employee, a group of employees, or a "division" of the company conducting the study, it cannot be someone, or group, that is otherwise part of the study. It can also be an "outside" individual or private firm or organization with experience in such matters hired by the company to perform the audit.

There are advantages to utilizing an "outside" individual or organization to perform the audit. First, such an arrangement provides company management and the study director with an obviously non-biased and fresh perspective on the operation and may rejuvenate a company's interest in the activity. It would also provide a third-party look at a company's operation and confirm with company officials areas that need improvements — especially if adherence to the GLP is at issue. Such an arrangement would also identify situations or infrastructure that are "weak links" in a chain: items that previously were overlooked due to the internal personnel being "too close to the action." One very important advantage is the fact that an audit by an outside individual or organization can prepare a laboratory well for an EPA or FDA audit, which is sure to follow at some point.

Facility Audit: The facility audit includes all aspects of the facility that are listed in the GLP regulations. These include the personnel (Subpart B) as well as the laboratory itself (Subpart C), the equipment used in the laboratory (Subpart D), the test substances, reagents, and samples used (Subpart F), and the laboratory information systems.

Concerning the personnel, the QAU would examine personnel files (Figure 8.1) to determine whether the education, training, and experience of the workers meet the requirements (see Section 7.2). It would also be interested in knowing if the information contained in the files is adequate and current. In addition, the QAU would want to interview the lab workers to determine if they have an adequate knowledge of the GLP regulations and how to apply them.

As to the actual lab facilities, the QAU would determine if there is adequate separation of activities, if the lab is neat and clean, if the waste disposal methods are adequate, and if the SOPs are current with respect to the work being performed.

As to equipment, the QAU would check to see if it is clean, well maintained, and regularly inspected, as would be evidenced by inspection tags and log books. The unit would want to know if the manuals and log books for equipment use, calibration, and maintenance are readily available and being used properly. The QAU would want to know if calibration is being done correctly and is properly documented, and whether all documentation is recorded in ink. Finally, the unit would check for proper documentation of errors, including initials, dates, and explanations, and whether changes are made so as not to obscure the original entry.

Figure 8.1 During a facility audit, the QAU examines all personnel files to see if educational requirements are met.

In terms of test substances, reagents, and samples, the QAU would want to look in the notebooks or logbooks to see if these substances are properly identified. It would also want to look at the container labels for proper identification (identity, concentration, composition, storage requirements, and expiration date). It would want to take note of expiration dates to see if all test substances and reagents are current. It would look for proper storage of substances, chain-of-custody documentation (date, quantity, distributor, receiver, etc.), and sample labels to see if they are adequate to avoid mixups.

Also included in a facility audit are the computer programs and systems used to generate data. Most important is whether or not these have been validated and tested for system suitability.

Conduct Audits: The QAU also inspects the manner in which the laboratory work is actually being carried out, the so-called "conduct audit," ostensibly again to observe compliance, or lack of compliance, with the GLP regulations. For this, the unit actually observes the lab worker doing his/her work. Important here would be the handling of test, control, and reference substances (Subpart F), compliance with the protocol used for the study (Subpart G), adherence to the SOPs used in the study and to the rules regarding deviations from SOPs (Subpart E), adherence to record-keeping rules (notebooks, logbooks, information management, etc.) and data handling techniques (Subpart J).

Final Report Audits: The QAU must sign off on the final report for the study. To do this, this unit must examine the final report as written and assembled by the study director and his/her subordinates. In the process, it inspects all records and raw data included in the report or influencing the presentation in the report. Thus, all original raw data, including notebooks, computer printouts, chromatograms, worksheets, chain-of-custody sheets, etc., are probed by the QAU. In addition, the protocol and SOPs are once again studied to confirm that the information in the report is truly derived from work that was carried out according to the protocol for the study and the approved SOPs. The unit looks at the analytical values reported (perhaps looking at a random sample if the number of such values is large), cross-checking these with those found in the study records. The unit also looks at the text of the report checking for technical errors, and cross-checking with the methods, equipment, results, and conclusions as presented in the study records.

Contract Lab Audits: Quality Assurance Audits also occur at the site of the contract laboratories. Of course, the contract laboratories must be aware from the beginning that their work will come under the scrutiny of the QAU. This means that laboratories must be informed that their work must comply with GLP and that these laboratories must make provisions for the QAU to come on site to examine the facilities, the equipment, the records, and the laboratory report to the sponsor. As stated previously, the QAU looks for the same things at the contract lab as in the company lab.

QAU Report: Following the audits described above, the QAU discusses the findings with the study director and writes an audit report. The study director responds to the QAU report in writing; the report is given to the management, signed, and included in the final report.

Commonly Found Errors: According to an American Chemical Society short course in which the concepts of GLP regulations and ISO 9000 accreditation are discussed, errors that the QAU commonly finds in the course of its work are mostly errors of record-keeping. The list provided at the short course is reproduced in Box 8.1.

Common Errors Found During Laboratory Audits

Misspelled words
Mathematical errors (rounding)
Wrong entry (e.g., data, sample number, etc.)
Entries not in sequence
Entries transcribed from other notes
Unauthorized procedural change
Wrong conclusion
Illegible entry
White-outs
Write-overs
Incomplete entry
No identification of individual making entry
No date of entry
No explanation/reason for change in entry

Box 8.1   Common errors found during laboratory audits according to the GLP/ISO9000 American Chemical Society short course. Used with permission.

8.2   EPA/FDA Audit

The EPA and FDA audit is a formal process involving advance notice, planning and scheduling (Figure 8.2), and a follow-up report, as well as the actual on-site observations and interviews. The advance notice includes an announcement of what specifically is to be audited. When the auditors arrive, the study director and lab workers are interviewed as a group, in what may be called an "opening conference," and the audit plan is laid out for everyone's full understanding. Included here would be a review of the purpose of the audit, the schedule to be followed, the scope of the audit, and the ground rules. The audit itself involves a formal schedule for the auditors, an inspection of the master schedule for the study, an inspection of the facilities and equipment, one-on-one interviews with the laboratory personnel, an examination of the study records (including the raw data, the study reports, and correspondence), the QAU procedures and reports, etc. There is a formal audit report and a follow-up. The auditors want to know if the data were generated properly, if the data support the conclusions of the

Figure 8.2 An EPA or FDA audit involves advance notice, planning, and scheduling.

study or the study reports, and whether the study is in compliance with existing protocols.

Any portion of the study that is a "field study" may also be audited. A field study auditor would inspect many of the same items already mentioned, but at the field site location. These include the training and experience of the field personnel, the calibration and maintenance of equipment, the field management and operations, the test substance application, and the sampling. Special problems are sometimes encountered in the field because the site is physically displaced from the main site of the study. Despite the physical displacement, the equipment must still be calibrated and maintained, the protocols and SOPs must be followed, there must be proper record-keeping, and there must be regular inspections by the QAU.

Audits can be stressful due to the fact that the consequences of significant violations discovered during audits can be severe. It is important to be prepared, to be courteous and forthright while the audit is being conducted, and maintain records of the audit and file an internal report. Some key elements of each of these critical components of an audit are presented here.

Preparation: In preparing for an audit, all data and documentation for the study are assembled for the inspectors to view (Figure 8.3). The study director, or other designated technical contact, should review the files and prepare to answer any questions. The master schedule and all files assembled by the quality assurance unit should be reviewed. An agenda, a work space, and a test facility floor plan should be prepared for the inspectors. The staff should be fully briefed and prepared for the visit.

Conducting the Audit: A staff person should accompany the auditors at all times so that there can be immediate attention to their needs and so that immediate corrective action can be taken on items the auditors point out. It

Figure 8.3   In preparing for an audit, all data for a study is assembled for inspectors to view.

is important that all staff persons be courteous and cooperative at all times and that all documents be immediately available in the work area. Any other items the auditors request should be provided. A daily debriefing session should be held to assess the progress of the audit. It is important to maintain a log of the activities and to also maintain a file of duplicate copies of all documents given to the auditors. Any proprietary information should be protected.

Completing the Audit: At the conclusion of the audit, an exit interview should be conducted so that all concerns can be immediately conveyed to staff. It is important to be aware of questions or concerns the auditors may have before they leave. A file of inspection documents should be maintained for an indefinite period after the auditors leave, and an audit report should be prepared.

9 Accreditation and certification

Accreditation refers to the formal approval of an organization through a prescribed process of inspection by an authorized body. The organization then is publicly recognized by this authorized body as fully capable of performing its function. Thus, a college in the Midwest, for example, applies to the North Central Association of Colleges and Schools, the body authorized to issue school accreditations in the Midwest, and, following a process of site inspections and interviews, becomes accredited by this organization. The public then has some confidence that the offerings and services of the college are up to quality standards.

As mentioned in Section 2, companies that manufacture products or provide a service for international utilization in some fashion pursue ISO 9000 certification. The steps involved include an internal review and the establishment of an internal program for quality, including a written **Quality Manual** outlining a **Quality Plan** with procedures to be implemented and maintained. An external assessment body reviews the manual and visits the company sites. Upon the assessment body's approval, the company becomes certified. Benefits include fewer corrective actions, a higher employee morale (Figure 9.1), improved internal and customer communications, better safety and housekeeping performance, enhanced product or service quality, and a lowering of costs. For an analytical laboratory operation within these companies, the ISO/IEC Guide 25 provides the general requirements to be followed. The GLP regulations should be followed. Items involving the laboratory that are audited during the ISO certification process include document control, product control, calibration and record-keeping, internal audit documents, how problems are resolved, and training.

Laboratories can become independently accredited. In this case, accreditation means that some independent auditor has reviewed the laboratory's staff, capabilities, and procedures against some set of requirements and standard practices and judged the laboratory and the workers in the laboratory capable of routinely performing some kind of laboratory work while adhering to those requirements and practices. Some accreditations are for a specific task or for a specific process; others are for laboratory work in general.

Figure 9.1　A benefit of certification can be a higher morale among employees.

Because laboratory testing often applies to articles in international trade, international accreditation is often sought. There are many international accrediting agencies in existence. Two examples are the National Institute for Standards and Technology (NIST) and the American Association for Laboratory Accreditation (AALA). NIST administers what is called the National Voluntary Laboratory Accreditation Program (NVLAP).

The NVLAP is comprised of a series of laboratory accreditation programs (LAPs), depending on what is requested and what is needed. Each LAP includes specific calibration and/or test standards and related methods and protocols assembled to satisfy the unique needs for accreditation in a field of testing or calibration. NVLAP accredits public and private laboratories based on evaluation of their technical qualifications and competence to carry out specific calibrations or tests. All are based on the ISO/IEC Guide 25 discussed in Section 2. The process involves an application and the payment of fees. This is followed by an on-site assessment. If deficiencies are found, these must be resolved to the satisfaction of the assessment team. The laboratory then participates in proficiency testing (Section 5.7) and a technical evaluation.

The AALA accreditation is also based on the ISO/IEC Guide 25. This accreditation indicates that a laboratory has demonstrated that it is competent to perform specific tests or specific types of tests; that its quality system addresses and conforms to all elements of the Guide 25 and is fully operational; and that it conforms to any additional requirements of AALA or specific fields of testing or programs necessary to meet particular user needs.

Bibliography

1. Gillis, J. and Callio, S., *Quality Assurance/Quality Control in the Analytical Testing Laboratory*, ACS Short Course Manual, American Chemical Society, Washington, D.C., 1997.
2. Mathre, O. and Schneider, P., *Good Laboratory Practices and ISO 9000 Standards: Quality Standards for Chemical Laboratories*, ACS Short Course Manual, American Chemical Society, Washington, D.C., 1996.
3. Prichard, E., *Quality in the Analytical Chemistry Laboratory*, John Wiley & Sons, New York, 1995.
4. Taylor, J.K., *Quality Assurance of Chemical Measurements*, Lewis Publishers, Boca Raton, FL, 1987.
5. *Good Laboratory Practice Regulations*, Part 58 (21CFR), Food and Drug Administration, Washington, D.C.
6. *Good Laboratory Practice Regulations*, Part 160 (40CFR), Environmental Protection Agency, Washington, D.C.
7. Huber, L., *Good Laboratory Practice and Current Good Manufacturing Practice*, Hewlett-Packard, Germany, 1993, 1994.
8. Anderson, M., *GLP Essentials: A Concise Guide to Good Laboratory Practice*, Interpharm Press, Inc., Buffalo Grove, IL, 1995.
9. Taylor, J.K., *Standard Reference Materials: Handbook for SRM Users*, NIST Special Publication 260-100, National Institute of Standards and Technology, Gaithersburg, MD, 1993.
10. Dux, J.P., *Handbook of Quality Assurance for the Analytical Chemistry Laboratory*, Van Nostrand Reinhold, 1986.
11. *Quality Systems — Model for Quality Assurance in Design, Development, Production, Installation, and Servicing*, ANSI/ASQC Q9001-9004, American Society for Quality Control, Milwaukee, WI, 1994.
12. Farrant, T., *Practical Statistics for the Analytical Chemist: A Bench Guide*, The Royal Society of Chemistry, Cambridge, England, 1997.
13. Harris, D.C., *Exploring Chemical Analysis*, W.H. Freeman and Company, New York, 1997.
14. Mookherjea, S., and Mathre O., *Classical and Wet Chemistry: Applications in Industrial and Pharmaceutical Analysis*, ACS Short Course Manual, American Chemical Society, Washington, D.C., 1998.
15. Youdon, W.J., *Experimentation and Measurement*, NIST Special Publication 672, Washington, D.C., 1961, Reprinted, 1997.

Homework exercises

1. Describe the principles of total quality management (TQM).
2. Describe the role that a chemistry laboratory worker or a chemistry process operator might have in implementing TQM.
3. What is meant by a "Quality System"?
4. What do the following acronyms stand for: ISO, ANSI, and ASQ? Describe the special role each of these organizations plays in assuring quality.
5. What is meant by ISO 9000? What is the ISO/IEC Guide 25?
6. What is meant by the following acronyms: cGMP, GLP, FDA, EPA, CFR?
7. What does 40 CFR 160 refer to?
8. Define: Quality Control, Quality Assurance, sample, analyte, validation study, accuracy, precision, bias, calibration, calibration curve, systematic error, determinate error, random error, indeterminate error, and outlier.
9. Given this data set, absorbance readings on an atomic absorption instrument,

Trial #	Absorbance	Trial #	Absorbance
1	0.672	26	0.676
2	0.673	27	0.679
3	0.680	28	0.675
4	0.675	29	0.673
5	0.677	30	0.665
6	0.685	31	0.671
7	0.675	32	0.678
8	0.676	33	0.681
9	0.671	34	0.676
10	0.662	35	0.669
11	0.674	36	0.676
12	0.679	37	0.671
13	0.678	38	0.677
14	0.673	39	0.679
15	0.675	40	0.672
16	0.679	41	0.675
17	0.674	42	0.681
18	0.670	43	0.676

Trial #	Absorbance	Trial #	Absorbance
19	0.676	44	0.669
20	0.668	45	0.670
21	0.677	46	0.680
22	0.674	47	0.674
23	0.682	48	0.679
24	0.678	49	0.672
25	0.676	50	0.682

determine the following statistically important items:

a. Mean, deviations, variance, standard deviation, RSD (percent and ppt).

b. If the warning limits are set at ±2 standard deviations, are any measurements outside the warning limits? If so, which? Show or tell how you determined this.

c. Establish confidence limits using the Student's t test. At the 95% confidence level, do any values lie outside the confidence limits?

d. Plot a histogram of these data, using ranges of 0.003 units (e.g., 0.674–0.676) on the x-axis and frequency of occurrence on the y-axis.

10. What is the significance of the following numbers with respect to the normal distribution curve: 68.3%, 95.5%, 99.7%?

11. Under what conditions are the mean, m, and the standard deviation, s, given the symbols μ and σ, respectively?

12. Consider the following results to an analysis: 359.2 ppm, 358.4 ppm, 361.9 ppm, 360.3 ppm, 354.5 ppm, and 359.7 ppm. Determine if any of these values should be rejected based on the Q test.

13. What is the meaning of each of the following acronyms: SOP, ASTM, and AOAC?

14. How does an SOP differ from an ordinary laboratory procedure found in an academic laboratory manual?

15. Check the Code of Federal Regulations, Chapter 40, Part 160.81, and tell whether an SOP is required for maintenance and calibration of equipment.

16. Write an SOP for the calibration of an analytical balance. Include a revision number, an indication of what SOP it replaces, the effective date, the purpose, the scope, the outline of the procedure, and references. Also include your name under "prepared by" and another name for "approved by."

17. Give an example of a calibration in which the "true response" for the device is already known. Give an example of one in which the "true response" needs to be established.

18. Give an example of a calibration in which the device readout is electronically adjusted to give the "true response." Give an example of a calibration in which the device readout cannot be electronically adjusted to give the "true response." In this latter case, how do you handle the situation?

19. What are the meanings of the following acronyms: RM, CRM, SRM, NIST?
20. What is the ultimate reference material and where can you obtain one?
21. What is meant by "traceability"?
22. What is the role and mission of NIST?
23. Define: statistical control, control chart, warning limits, action limits.
24. Imagine that a particular laboratory checks the calibration of an analytical balance on a daily basis with a set of known weights. The following weights were obtained for a known weight of 5.0000 grams. Construct a quality control chart of these data assuming that 5.0000 is the desired value.

Day 1 - 5.0001   Day 2 - 5.0000   Day 3 - 4.9998   Day 4 - 5.0000
Day 5 - 4.9999   Day 6 - 5.0002   Day 7 - 5.0000   Day 8 - 5.0001
Day 9 - 5.0004   Day 10 - 5.0003   Day 11 - 4.9999   Day 12 - 5.0001
Day 13 - 5.0004   Day 14 - 5.0004   Day 15 - 5.0005   Day 16 - 5.0004
Day 17 - 5.0006   Day 18 - 5.0004   Day 19 - 5.0001   Day 20 - 5.0004

Assuming a standard deviation of 0.0002, add warning limits at ±2 standard deviations and action limits at ±3 standard deviations to your control chart. Is there any day that you would take the balance out of service and perhaps call in a service agent? Explain.

25. Define: noise, blank, detection limit, signal-to-noise ratio, useful range, ruggedness, selectivity.
26. Explore the ASTM Web site (www.astm.org) and navigate to the ASTM store. Examine the list of standard methods that are available for viscosity measurements. Report on the cost and description of one of them.
27. Locate a copy of the USP National Formulary and report on a method for determining the active ingredient in a pharmaceutical product.
28. Define: validation, system suitability, proficiency testing.
29. Explain why establishing statistical control for a new method is so involved.
30. Define: sample, representative sample, composite sample, selective sample, random sample, bulk sample, primary sample, secondary sample, subsample, laboratory sample, test sample.
31. Explain why statistics is important for sampling. Compare your reasons to the reasons why statistics is important for lab analysis.
32. Define chain of custody. Why is it critical to document the chain of custody?
33. What two federal agencies have GLP regulations? <u>Briefly</u> describe their similarities and differences.
34. Define the following terms from the GLP regulations: quality assurance unit, raw data, person, sponsor, study, study director, test system.
35. What statements in the GLP regulations give the government the right to enter and inspect a laboratory facility?

36. What are the duties of the managers of a test facility according to GLP?
37. What is the purpose of the quality assurance unit according to GLP regulations?
38. Can a test system, according to the GLP regulations, be an animal? Explain.
39. How is "protocol" defined according to GLP regulations? Describe its importance.
40. What are two entities that are permitted to audit a laboratory facility? What are some of the things that auditors look for?
41. How does a laboratory prepare for a government audit? How is an audit conducted? What happens at the conclusion of an audit?
42. What is meant by the following acronyms: AALA, NVLAP?
43. How does a laboratory facility become accredited? What role does the ISO/IEC Guide 25 play in accreditation by AALA or NVLAP?

Index

A

AALA, See American Association for
    Laboratory Accreditation
Accreditation and certification, 69-70
Accuracy, 8, 36, 38, 39
Action Limits, 33, 34, 35
American Association for Laboratory
    Accreditation (AALA), 70
American National Standards Institute
    (ANSI), 5, 6
American Society for Quality (ASQ), 5
American Society for Testing and Materials
    (ASTM), 21, 22, 39
Analyte, 8
Analytical balance, 26, 27, 36
Analytical chemistry laboratory, 1, 3
ANSI, *see* American National Standards
    Institute
AOAC International, *see* Association of
    Official Analytical Chemists
ASQ, *see* American Society for Quality
Association of Official Analytical Chemists
    (AOAC), 21, 22, 39, 40
ASTM, *see* American Society for Testing and
    Materials
Atomic absorption spectrophotometry, 30, 37
Audits, 51, 63-68
    Quality assurance, 63-66
    EPA/FDA, 64, 66-68
    Facility, 64-65
    Conduct, 65
    Final Report, 65
    Contract lab, 65
Auto-titrators, 27

B

Balance, 9
Bias, 8, 11, 15, 17, 18, 20, 33, 34, 36, 38, 39, 59
Blank, 37
Buffer solutions, 28
Bulk sample, *see* Sample
Burets, 25

C

Calibration, 8, 9, 11, 21, 22, 24-31, 32, 34, 35,
    56, 64, 67, 69, 70
    Certificate, *see* Certificate of calibration
    Constant, 29
    (standard) curve, 9, 25, 30, 32, 41
Cannon-Fenske Tube, 28, 29
Capacity (of a laboratory), 36, 38, 39
Capillary electrophoresis, 36, 38
Certification, *see* Accreditation
Certified Reference Material (CRM), 31
CFR, *see* Code of Federal Regulations
cGMP, *see* Current Good Manufacturing
    Practices
Chain of custody, 48, 58, 65
Chromatographs, 37
    Gas (GC), 30-31
    Liquid (HPLC), 30-31
Class A glassware, 25
Code of Federal Regulations (CFR), 6, 21, 22,
    24, 33, 50
Coefficient of variance, 14
Compliance/non-compliance, 51, 53, 65
Composite sample, *see* Sample
Conduct audit, *see* Audit
Confidence level, 17
Confidence limits, 16, 17, 32
Confidence interval, 17, 18

Milton Keynes UK
Ingram Content Group UK Ltd.
UKHW031152141024
449569UK00024B/850